Amazonian Geographies

Amazonia exists in our imagination as well as on the ground. It is a myster-ious and powerful construct in our psyches yet shares multiple (trans)national borders and diverse ecological and cultural landscapes. It is often presented as a seemingly homogeneous place: a lush tropical jungle teeming with exotic wildlife and plant diversity, as well as the various indigenous populations that inhabit the region. Yet, since Conquest, Amazonia has been linked to the global market and, after a long and varied history of colonization and devel-opment projects, Amazonia is peopled by many distinct cultural groups who remain largely invisible to the outside world despite their increasing integra-tion into global markets and global politics. Millions of rubber tappers, neo-native groups, peasants, river dwellers, and urban residents continue to shape and re-shape the cultural landscape as they adapt their livelihood practices and political strategies in response to changing markets and shifting linkages with political and economic actors at local, regional, national, and interna-tional levels.

This book explores the diversity of changing identities and cultural land-scapes emerging in different corners of this rapidly changing region.

This book was published as a special issue of the *Journal of Cultural Geo-graphy*.

Jacqueline M. Vadjunec is an Assistant Professor of Geography at Oklahoma State University.

Marianne Schmink is Professor of Latin American Studies and Anthropology at the University of Florida, where she was Director of the Tropical Con-servation and Development (TCD) program.

Amazonian Geographies
Emerging Identities and Landscapes

Edited by
Jacqueline M. Vadjunec and
Marianne Schmink

Routledge
Taylor & Francis Group

LONDON AND NEW YORK

First published 2012
by Routledge
2 Park Square, Milton Park, Abingdon, Oxfordshire OX14 4RN

Simultaneously published in the USA and Canada
by Routledge
711 Third Avenue, New York, NY 10017

First issued in paperback 2014

Routledge is an imprint of the Taylor and Francis Group, an informa business

This book is a reproduction of the *Journal of Cultural Geography*, vol. 28, issue 1. The Publisher requests to those authors who may be citing this book to state, also, the bibliographical details of the special issue on which the book was based.

Trademark notice: Product or corporate names may be trademarks or registered trademarks, and are used only for identification and explanation without intent to infringe.

British Library Cataloguing in Publication Data
A catalogue record for this book is available from the British Library

ISBN 978-0-415-60053-8 (hbk)
ISBN 978-1-138-79834-2 (pbk)

Typeset in Times New Roman
by Taylor & Francis Books

Disclaimer
The publisher would like to make readers aware that the chapters in this book are referred to as articles as they had been in the special issue. The publisher accepts responsibility for any inconsistencies that may have arisen in the course of preparing this volume for print

This book is dedicated in loving memory of Vanessa Sequeira

Contents

New Amazonian geographies: emerging identities and landscapes

Jacqueline M. Vadjunec[a], Marianne Schmink[b] and Alyson L. Greiner[c]

[a]Department of Geography, Oklahoma State University, Stillwater, OK, USA;
[b]Center for Latin American Studies, University of Florida, Gainesville, FL, USA;
[c]Department of Geography, Oklahoma State University, Stillwater, OK, USA

Common stereotypes of a homogeneous Amazonia belie the complexity and diversity of peoples and landscapes across the region. Although often invisible to the outside world, diverse peoples—indigenous, traditional, migrant, urban dwellers and others—actively construct their identities and shape cultural and political landscapes in diverse ways throughout the region. This volume combines political ecology, with its emphasis on identity, politics, and social movements, with insights from cultural geography's focus on landscapes, identities and livelihoods, to explore the changing nature of Amazonian development. These papers focus on indigenous identity and cosmology; changing livelihoods and identities; and transboundary landscapes. They highlight the diversity of proactive, place-based social and political actors who increasingly raise their voices to contest and engage with Amazon development policies. Based on their history, social values, and livelihood practices, such groups propose alternative ways of understanding and managing Amazonian landscapes.

"Gigantic," "green," "emerald," "wet," "humid," "important," "(bio)-diverse," "lungs of the earth," "enormous," "in danger" and "full of endangered species." Amazonian researchers are accustomed to hearing people from many backgrounds—ranging from young schoolchildren, to university students, to other citizens and educators—use words like these to describe Amazonia. Such lush, larger than life perceptions of Amazonia dominate the mental landscape of those not familiar with the region. Less often, or after a few minutes of conversation, the description sometimes turns to talk about the indigenous peoples who reside in the region and "live in harmony with nature." While these vivid images of Amazonia might help to fix the region in the imagination of the general public, the reality of Amazonia is both more complex and diverse.

Amazonia is a mysterious and powerful construct in our psyches, yet shares all-too-real (trans)national borders and diverse ecological and cultural landscapes. It is often presented as a seemingly homogeneous place: a lush tropical jungle teeming with wildlife and plants, as well as timeless Indians. Rarely are outsiders aware of the immense diversity of Amazonian flora and fauna, or of the fantastic stories of the varied peoples who inhabit the different corners of the region. As a result, Slater (2002, p. 203) argues that it is time for us to move "beyond Eden" and re-envision an Amazon that encompasses the diverse groups of people who live there, and the complexities of their interactions with one another and with the natural environments of their territories.

Far from a pristine jungle, Amazonia has since Conquest been linked to the world through global markets. After a long and varied history of migration, colonization, and development projects, Amazonia is peopled by many distinct and "other" cultural groups who are still invisible to the outside world despite their increasing integration into global markets and global politics. Millions of rubber tappers, neo-native groups, peasants, river dwellers, and urban residents continue to shape and re-shape the cultural landscape. They adapt their livelihood practices and political strategies in response to changing markets, and to shifting linkages with political and economic actors at local, regional, national, and international levels. This volume explores the diversity of changing identities of those inhabiting the region, and of the cultural and political landscapes they are constructing in different corners of this rapidly changing region today. It also traces how Amazonian groups draw on their place-based history, social values, and livelihood practices to challenge dominant development paradigms and propose alternatives more suited to their identities and aspirations.

Carving out Amazonian geographies: contested spaces and changing landscapes

The Amazon, the second largest river in the world, flows approximately 2,320 miles from just outside Iquitos, Peru, though parts of Colombia, to the Atlantic Ocean near Macapá, Brazil (Goulding *et al.* 2003). The Amazon River basin extends over 2.5 million square miles, the majority of which is covered with tropical rainforest (London and Kelly 2007). Amazon rainforest covers much of Brazil, Peru, Bolivia, Colombia, Ecuador, Venezuela and Guyana, Suriname and French Guiana (Goulding *et al.* 2003).

The Amazon region is rich in a wide variety of minerals, including one of the largest gold reserves in the world (Hecht and Cockburn

1990; Schmink and Wood 1992). It has greater plant and animal diversity than anywhere else on earth: an estimated one million plant and animal species (20% of all total species on earth) inhabit the region (London and Kelly 2007). The extractive potential of this region has long been looked upon as seemingly endless—fruit, nuts, timber, rubber, medicinals, and minerals are just a few of the many natural resources that dominate the region. Furthermore, the Amazon watershed provides the largest source of fresh water on the planet (Goulding *et al.* 2003). This important ecosystem, however, is also a hotspot for tropical deforestation (Skole *et al.* 1994; Achard *et al.* 2002).

Since its "discovery" by Europeans, people have been fighting over what Hecht (2004) refers to as the "mythic," empty Amazon. In 1494, the Tordesillas Treaty was signed between Spain and Portugal, granting Portugal everything 370 leagues to the west (Hecht and Cockburn 1990; Roux 2001). This gave much of what is now Brazil to the Portuguese. Early on, the French, the Dutch, and the Germans entered Amazonia along the eastern coast through the Guianas, with the goal of establishing trading posts and colonies (Hecht & Cockburn 1990). Additionally, the Spanish initially explored the Amazon region spurred by dreams of El Dorado, the lost city of gold. Increased exploration of the region in the eighteenth and nineteenth centuries, the rubber boom (Dean 2002; Stokes 2000), poor mapping, treaty disputes, and other "cartographic uncertainties" (Hecht, forthcoming; Salisbury *et al.* 2011) further proved the need for countries to secure their own boundaries, while carving up and parceling out the vast territories of the Amazon region.

While each country has a unique history of Amazon development trajectories (see Hecht, forthcoming), many decisions regarding how, when, and where to develop the Amazon have been encouraged by the perceived need of filling the "demographic void" (Hecht and Cockburn 1990). In Brazil, for instance, whose developmental history is perhaps the most extensive, the military government's geopolitical quest to fill the empty spaces of Amazonia and *"integrar para não entregar"* (integrate in order not to forfeit), led to General Medicí's (1969–1974) Plan for National Integration (PIN) (Treece 1994, p. 62). This plan sought to achieve Brazil's economic miracle through the "rational use" of the Amazon's resources (Guimâraes 1991). Medicí's development plan also attempted to address social inequities by encouraging the migration of Brazil's rural poor to the Brazilian Amazon, in an attempt to unite "men without land with land without men" (Schmink and Wood 1992, p. 105; Hall 1997, p. 47). The Plan for National Integration led to massive colonization schemes, known as Integrated Colonization Projects (PICs), and other large-scale agrarian reform programs known as Directed Settlement Projects (PADs), all requiring large-scale infrastructure, resulting in "big development" projects such as the building of the Transamazon Highway (Moran 1981; Smith 1982; Ludewigs 2006).

By the 1970s, Brazilian policy shifted once again, away from the smallholder colonists and towards large-scale cattle ranching and capital intensive mega-projects. The military government and the National Security Council provided ranchers with generous incentives, such as tax breaks and highly subsidized credit (Schmink and Wood 1992; Hecht 1993). Between 1971 and 1987, these incentives amounted to over 5.15 billion dollars (Hall 1997, p. 50). According to Fearnside (1997, p. 549), as much as 70 percent of historical deforestation in the Brazilian Amazon has been caused by medium and large-scale cattle ranchers. Such colonization projects and large-scale investments continue throughout Amazonia from Puerto Maldonado, Peru to the upper Napo Basin in Ecuador (Goulding *et al.* 2003; Perz *et al.* 2005).

The loosening up of policies after the fall of many of the military governments in the 1970s and 1980s was followed in the 1990s by the expansion of neoliberal policies favoring large infrastructure projects to support export and trade, such as those described by Pieck (2011). In response to the impacts of such development policies, and due to the growing influence of worker's unions (Keck 1995), liberation theology and the Catholic Church (Freire 1970), along with growing international concerns by environmentalists regarding the destruction of the Amazon rainforest, new social movements formed, pressing for both environmental and social justice throughout Amazonia (Allegretti 1990; Hecht and Cockburn 1990). These networks of allied social movements, and the local Amazonian communities tied to them, not only contested but also engaged with the state to construct alternative discourses and practical proposals for development, based on values of citizenship, participation, community, justice, and the moral economy.

Far from a demographic void, by the time of rubber tapper and union leader Chico Mendes' assassination by cattle ranchers in 1988 (Revkin 2004), Amazonia was a bustling frontier inhabited by a diversity of peoples—native, traditional, resurgent indigenous, palm-nut breakers, migrant colonists, and, increasingly, urban dwellers—including NGO and government functionaries as well as ranchers, loggers, and bankers. These different groups were pitted against one another in struggles to support or contest current development models as they were implemented across already-complex social and ecological landscapes. The papers in this volume highlight how such struggles have taken on nuanced and shifting forms, from contestation and resistance, to strategic engagement with the state and allies at diverse scales.

Amazonia today: diverse peoples and changing identities

Today, Amazonia is home to over 20 million people, a number that remains unfathomable to most outsiders (Slater 2002, p. 10). This figure includes both urban and rural dwellers, ranchers, loggers, commercial farmers, as well as a diverse array of traditional communities and social movements (Almeida 2008). The New Social Cartography Project of Amazonia (PNCSA), led by anthropologist Alfredo Wagner de Almeida, has systematically mapped and documented territories occupied by indigenous peoples (approximately 200,000–300,000 people), *quilombolas* (former slave groups—perhaps 2 million people), palm-nut breakers (approximately 400,000 people), as well as uncounted forest extractivists, fishers, and dozens of other self-defined communities (Almeida 2008; http://www.novacartografiasocial.com). These emergent communities are collective organizations created to represent common interests and press for legal recognition of traditional forms of land occupation and resource use (Almeida 2008). The multiple processes of territorialization and identity construction blur the lines among the major Amazonian social groups, as briefly described below.

Indigenous peoples

Cultural development of indigenous Amazonian peoples was long thought to be limited by soil and climate deficiencies that made the apparently luxuriant jungle a "counterfeit Paradise" (Meggers 1971), but more recent research has shown how those native groups produced rich, anthropogenic *terras pretas* or "black earths," and their pottery shards reveal that the Amazon has been populated, in some places quite densely, for at least ten thousand years (Denevan 1996; Smith 1999; Glasner and Woods 2004). A growing body of archaeological and linguistic research suggests that much of the current Amazon forest consists of "cultural landscapes," the product of widespread and long-term anthropogenic manipulation, unlike previous models of pre-Colombian Amazonian peoples that conceived of them as living in small, dispersed settlements (Balée 1994; Heckenberger *et al.* 2007, 2008).

Smith (1999, p. 28) estimates that the native population of Amazonia was approximately 15 million at the time of European contact in 1500, consisting of hundreds of tribes. Today, over 200 native tribes remain. These tribes range in size from 200 to 30,000 people and their members speak over 180 languages from 30 language families (http://pib.socioambiental. org/pt/c/no-brasil-atual/quem-sao/introducao). Several isolated tribes have little or no contact with state governments or the broader society. In Brazil, the National Indian Foundation (FUNAI) estimates over 50 such isolated tribes in the Brazilian Amazon, and others have been identified in Peru, Colombia, and Ecuador (Roach 2003).

Researchers and activists in Amazonia have long defended indigenous territories based on both social and environmental justice concerns (Ramos 1998; Hecht and Cockburn 1990; Colchester 2000; Cunha and Almeida 2000; Posey 2001). Beginning in the 1970s, Amazonian indigenous movements began to emerge, and to forge partnerships with international advocacy and environmental groups, achieving some successes in securing land and other rights (Fisher 1994; Conklin and Graham 1995; COICA 1996; Ramos 1998; Langer and Muñoz 2003). In Brazil, over one million km^2 of protected indigenous lands already exist, serving as a barrier to both deforestation and the spread of agricultural fires; these protected indigenous lands also promote cultural survival and simultaneously provide a means for protecting biodiversity (Nepstad et al. 2006, p. 66). While generally having greater land security in Brazil, many native groups in Peru, Ecuador, and Venezuela, continue to be involved in violent encounters and territorial disputes with gold miners, loggers, and oil companies (Goulding et al. 2003; Pieck and Moog 2009; Salisbury et al., 2011). Moreover, under threat from these same outside forces, many descendants of indigenous groups in Brazil who lost their language and other ethnic "markers" have begun to recuperate their indigenous past—a process known as "ethnogenesis"—as a means to gain land rights as well as cultural and political visibility (Bolaños, 2011). These emergent and shifting identities among indigenous groups are a reminder of the perennial territorial struggles that continue to shape identities and practices in the Amazon region (Little 2001).

Caboclos and Ribeirinhos

Caboclos, traditional Amazonian peasants, and *ribeirinhos*, or river dwellers, are broad categories falling into an often pejoratively perceived "other" class, generally having negative connotations and referring to disparate groups of so-called deculturated natives, people of mixed indigenous, European, and Afro-Brazilian descent, as well as pre-Transamazon Highway smallholder migrants (Parker 1985; Nugent 1993). Such social categories as rubber tappers, Brazil nut collectors, and babassu palm-nut breakers represent specific occupational specializations within the categories of *caboclo* or *ribeirinho*. *Caboclos* generally operate as peasant swidden farmers, practicing both subsistence and overflow market agriculture, in combination with their long historic ties to the market through non-timber forest products such as Brazil nuts and rubber collection. *Ribeirinhos*, on the other hand, make their living mainly from traditional fishing practices for both home and market consumption (Nugent 1993; Harris 2000). The livelihood systems of these groups in Amazonia's highly-seasonal conditions depend on a moral economy that links domestic units, binding people together through kin and neighbor ties, and providing a safety net during lean periods (Scott 1976; Harris 2000; Minzenberg and Wallace 2011). Although an uncounted number of *caboclos* and *ribeirinhos* live in the rural areas of Amazonia today, as Nugent (1993) argues, historically they are highly invisible in the Amazonian landscape and imagination of the outside world, neither ethnically distinct indigenous peoples nor modern urban citizens.

In the past 30 years, many of these Amazonian peasants, like indigenous groups, also have emerged from the shadows due to their successful political organizing (see Vadjunec *et al.* 2011; Porro *et al.* 2011). In the 1970s and 1980s these movements largely organized around land struggles and resistance to externally-imposed development programs, and over time they developed important alliances both inside and outside of Brazil that allowed them to engage directly as players in important policy debates. Like former slave communities, called *quilombolas*, these groups have taken advantage of the rights granted to "traditional peoples" in Brazil's 1988 Constitution, to press for their rights to land and resources, construct new political identities, and propose new re-source tenure regimes more compatible with their traditional norms and practices. In the absence of such political mobilization, and lacking governmental support for forest-based livelihoods practiced by *caboclos* and *ribeirinhos*, many of these traditional Amazonian peasants have shifted their livelihoods to agriculture and, primarily, cattle raising (Salisbury and Schmink 2007). Some have migrated to urban areas, while others have sought to reclaim their indigenous identities.

Migrant colonists

Spontaneous migration from other regions into the Amazon, historically associated with resource booms and expansion of the agricultural frontier, was eclipsed by state-directed development programs starting in the 1970s. Government-run colonization projects, big infrastructure projects, cattle ranching and Amazonian resource extraction such as gold mining and logging stimulated migration that led to rapid population growth in the Amazon region. Colonization schemes not only served hard-hit regions such as the northeast of Brazil, acting as a safety valve in times of drought, they also advanced the military government's geopoliti-cal strategies (Hecht and Cockburn 1990; Schmink and Wood 1992). The majority of colonists originally migrated to rural areas, but many have long since moved to urban centers, where new generations of Amazonians now are being born. For instance, the Transamazon High-way scheme implemented in the 1970s was a government-planned project that was supposed to settle one million families by 1980; however, less than 50,000 families currently remain (Goulding *et al.* 2003). In Acre, the southwesternmost Amazonian state of Brazil almost three times the population lived in rural areas than in cities in the 1970s; such trends reversed dramatically throughout Amazonia in the 1980s and 1990s (Schmink and Cordeiro 2008, p. 30). Some settler families, with favorable labor and natural resource endowments, have managed to stabilize and expand their production systems, while others have been forced to sell out, leading to consolidation of land, often in the hands of urban dwellers. Like *caboclos* and *ribeirinhos,* many of these settlers are shifting their productive focus to short-term strategies such as small-scale cattle-raising (Browder 1994; Almeida and Campari 1995; Walker *et al.* 2000).

Urban dwellers

Far from the wild Amazon found among the glossy, vivid pages of *National Geographic*, today's Amazon is highly urbanized. According to Goulding and colleagues (2003) as much as 75 percent of the lowland Amazon Basin population lives in cities. In Brazil, the percentage of the Amazonian population living in cities more than doubled between 1940 and 1991, reaching just under 60 percent (Browder and Godfrey 1997). The two major metropolitan areas of Belém and Manaus have grown rapidly, and are now home to over one million inhabitants each. Many of these urban dwellers are former *caboclos*, indigenous peoples, colonists, and migrants from rural areas, or their offspring. Godfrey and Browder argue that frontier settlement patterns are highly dynamic and fluid, thus making traditional, "established dichotomous categories of rural and urban ... problematic" (1997, p. 14), as rural dwellers increasingly combine rural livelihoods with urban residences, where many rural customs may persist in the towns and cities of the region. Such a diverse, highly dynamic, and rapid melding of Amazonian residents throughout a highly-contested region has produced emerging and complex identities as well as rapid changes in Amazonian cultural and political landscapes. The papers in this volume address changing transboundary landscapes, and fluid and shifting livelihoods and identities as emergent groups negotiate their place in Amazonian policies and practices.

Bridging identities, landscapes and livelihoods: political ecology and cultural geography

Political ecology — complex identities and social movements

Political ecology (PE), which emerged in the 1980s as a subfield in both geography and anthropology, is most commonly defined as an approach to land degradation that combines "concerns of ecology and a broadly defined political economy" (Blaikie and Brookfield 1987, p. 17). PE has since expanded and serves as an "umbrella term" that incorporates mainly non-mainstream science, political, institutional, and economic perspectives on social and environmental justice issues. Numerous variations of political ecology exist, including "Third World political ecology," which focuses on processes of social, political, and economic conflict and environmental change occurring in developing countries (Bryant and Bailey 1997); "Liberation Ecology," which calls for a poststructuralist political ecological approach that "integrates politics more centrally," (Peet and Watts 1996, p. 3); and "Feminist political ecology," which argues that gender is a key factor influencing resource access, control, and environmental change (Rocheleau *et al.* 1996, p. 28).

The political ecology of development in Latin America provides an essential backdrop for the emergence of environmental social movements, shifting landscapes and boundaries, and dynamic social and political identities in the Amazon region that are addressed in this volume. (See Robbins 2004 for a thorough discussion on identity and social movements in PE.) The chaotic, often violent, and rather rocky transition between the end of militarization and the birth of new democracies in Latin America, and the growing dominance of neoliberal development projects based on infrastructural development and export crops, led to resistance by Amazonian groups to the social and environmental impacts of these policies on their livelihoods and territories. Growing concerns over the negative social and environmental impacts of development, such as tropical deforestation, by local Amazonian populations as well as powerful international interests and NGOs, gave rise to new social movements, which increasingly adopted green discourses while focusing on pressing livelihood issues (Hecht and Cockburn 1990; Schmink and Wood 1992). The evolution from a red (socialist/workers' rights) discourse to a green (environmental) discourse brought increased power and mobility of historically marginalized groups, allowing them to reinvent themselves and gain new voice, while connecting diverse actors across both time and space. As Robbins (2004, pp. 188–189) explains: "such movements often represent a new form of political action, since their ecological strands connect disparate groups, across class, ethnicity, and gender."

New social movements are made up of diverse actors with complex, multiple, overlapping, and even conflicting reasons for participating (Rocheleau *et al.* 1996; Robbins 2004). In order to deal with issues of marginalization and environmental degradation, these new social movements may form from alliances based in part on gender (Rocheleau *et al.* 1996), race (Miller *et al.* 1996), shared experiences (Harris 2007), and/or ethnicity (Bebbington 2001), among other things. Furthermore, in light of globalization, many new social movements have the ability to jump scales, connecting local people directly to international NGOs (often bypassing the state in the process) in order to gain both power and legitimacy (Bebbington and Batterbury 2001; Heynen *et al.* 2006). Perreault (2001) illustrates how indigenous groups in Ecuador adopt and exploit indigenous identity and discourse in order to gain political power and access to resources. Escobar and Paulson (2005) show how a "reaffirmation of identity" is a key organizing principle in the fight over territorial rights among ethnic communities in Colombia.

While many of these new social movements initially gain power through the adoption of green discourse or through their historic, traditional, low-impact livelihood activities (a declaration of their environmental identity), they are also increasingly susceptible to growing criticism by environmentalists and policy makers when they or their land-use and livelihood activities are no longer perceived as green or traditional by outside supporters. As such, their very identity and legitimacy is often brought into question. Situating identity in relation to land-use and livelihood is therefore paramount to understanding complex environmental problems. Jackson and Warren (2005, p. 561) argue that identity construction needs to be analyzed as a process that is "fluidly multiple and ... relational." Understanding social movements and identity in the face of changing landscapes and livelihoods is of growing importance, particularly in rapidly changing regions of the world such as the Amazon.

One limitation of political ecology is that it often ignores the continuous "dialog" between human beings and their environment (Cronon 1994), therefore failing to consider environmental feedbacks. For instance, Black (1990, p. 44, original emphasis) argues that political ecologists mainly focus on one-way causality, "where it is argued that the state affects agrarian society, which in turn affects land management, but not *vice versa.*" Many of the articles in this volume illustrate how important local social actors can be in actively changing not only their own identities, but also the political and environmental landscapes of Amazonia. Organizing themselves to participate in resistance to or debates about infrastructural investments such as roads and dams and to propose alternative forms of land tenure and resource management rules and institutions, Amazonian groups both contest and engage with state development actors to negotiate the terms of their involvement with development.

Cultural geography — landscape, identity and livelihoods

Cultural geography, a wide-ranging, diverse and well-established subfield of geography, broadly claims culture and identity, in terms of both space and place, among its main subjects of study (Hugill and Foote 1994). Cultural geography's rich history and focus on identity make it complementary to the PE research on environmental identities and social movements. An important component of cultural geography involves the study of cultural landscapes and their change over time. In the process of landscape production, as Jackson (1989, p. 48) argues, history and geography are not passive agents, but "are actively forged by real men and women." In this way, landscapes are produced and constructed, serving as symbols of culture for society at large, and intimately entangled with one's identity (Olwig 2001).

Cultural geography also focuses on the differences of individual perceptions of and experiences in landscape and the connections between place, politics, culture, and identity (Keith and Pile 1993; Lowenthal

2001). Often this involves research on emotional geographies (Davidson *et al.* 2005), and attachment to, or sense of place (Tuan 1974; Duncan and Duncan 2004). Such studies highlight the importance of everyday lived experiences (Nash 2000; Ley 2001), and enable cultural geographers to trace the contours of what J. E. Malpas (2007, p. 174) calls "a philosophical topography." More specifically, "the nature and identity of individual persons in particular, is to be understood only in relation to place, and in relation to the particular places in which the subject is embedded" (Malpas 2007, p. 174).

Studies of place and identity are not limited to the individual scale, however. Equally important are the powerful emotional and affective bonds that form between groups of people, and certain sites, territories, and other spaces. In Amazonia, the process of sustaining these bonds has, arguably, never been more complicated or as highly politicized. One way to understand group identity, according to Alvarez *et al.* (1998), is to see it as a form of cultural politics. Group identity, then, while always fluid and contingent on the negotiations and struggles between different political, economic, and institutional actors, is ultimately about making meaning. In this light, discourse—understood as both practice and narrative—becomes a powerful way of defining values. At the center of major debates are discourses of "development," "traditional people," "indigenous," and "primary vs. secondary forest," among others. The papers in this volume contribute to our understanding of the fluid, fractal nature of diversely constructed and reconstructed cultural and political identities in Amazonia. They also highlight the importance of discourse and the relatively invisible cultural values and practices on which traditional Amazonian groups base their alternative proposals to the dominant neoliberal development model.

Today's most critical environmental problems such as tropical deforestation and global environmental change, linked with social justice concerns for developing countries, require "bridging" between research cores both within and beyond our discipline (Turner 1997, p. 199). This collection of papers echoes Turner and Robbins' (2008, p. 295) recent call for the combined use of "complementary but parallel approaches" of land-change science (LCS) and PE approaches in sustainability research. To these "hybrid" ecologies (Zimmerer 2006; Turner and Robbins 2008), which often incorporate a mixing of methods and even paradigms, we add a call for the rich detail and intricate understandings that can arise from studying the complex interactions between culture, place, and identity (see Bebbington and Batterbury 2001). Seriously engaging with identity will become increasingly important if we are to speak to the natural and applied sciences, policy-makers, governments, NGOs and environmentalists regarding the challenges and solutions for diverse Amazonian actors in the face of global environmental change.

11

New Amazonian geographies: emerging identities and landscapes

The papers in this volume shine a spotlight on some of the hitherto relatively obscure dimensions of the emerging new cultural and political landscapes of Amazonia. These papers, presented by geographers, anthropoligists, and a range of academic and independent scholars from both North and South America, focus on indigenous identity and cosmology; changing livelihoods and identities; and transboundary landscapes. They show the complexity of meanings associated with political boundaries and natural landscapes, and the evolving negotiations between local communities and the state, ranging from resistance to engagement. Based on increasingly fluid and fractal identities, these groups advance their alternative development proposals associated with distinct values, symbols, and memories tied to specific territories and lifeways.

The complexities of identity are explored in greater depth in the volume's first pair of papers on indigenous identity and cosmology. Anthropologist Laura Mentore explores the clash of discourse, and of underlying cosmographies, between the "arboreal, unilinear, commodified" framework of government development plans in Guayana, and the "fractal, recursive, mediated" cosmography of the Waiwai peoples. Ethnographic understanding shows that concepts such as "tree" and "river," embedded in memory, social relationships, attachment to place, and spirituality, are not readily translatable in Waiwai culture, much less the abstract notion of "Payments for Environmental Services (PES)"— even though such policy proposals are regularly assumed to be both universal and subject to straightforward translation.

Colombian anthropologist Omaira Bolaños takes up the issue of the revitalization of indigenous ethnic identities and territorial rights, and examines the ethno-political actions of the Indigenous Council of the Lower Tapajós-Arapiuns (CITA) to redefine themselves as indigenous peoples with stronger rights to their territories. These actions have triggered contesting discourses by non-indigenous actors who use essentialized notions of "indigenous" to question the legitimacy of these "resurgent ethnicities," in the same way that pristine notions of "wilderness" can be used to justify the expansion of deforestation in altered areas, considered to have low biodiversity values. These discourses serve to justify both the denial of indigenous land rights, and the continued expansion of agriculture into secondary forests.

The next three papers explore more deeply the complex links between changing livelihoods and changing identities among diverse non-indigenous Amazonian peoples: rubber tappers, *caboclos,* and babassu nut breakers. In their paper, geographer Jacqueline Vadjunec, anthropologist Marianne Schmink, and Brazilian geographer Carlos Valério Gomes trace the remarkable changes in livelihood practice, rural versus urban residence, and social identities among both rural and urban dwellers in the southwestern Amazonian state of Acre. Drawing on long-term fieldwork, they explore how the history of the rubber tappers' social movement, and its success in land rights conquests and influence on public policies for the Amazon, made the state of Acre a laboratory for alternative socio-environmental development models based on *Florestania:* "citizenship with a forest face." As landscapes have been shaped and reshaped by changing resource use practices, new and complex hybrid landscapes and practices are emerging, linked to increasingly fractal and complex identities among rubber tappers and city dwellers.

The paper by anthropologists Eric Minzenberg and Richard Wallace focuses attention on the importance of intangible, social aspects of livelihoods among traditional riverside-dwelling *caboclos* in the western portion of the state of Acre. Their analysis of the drivers of hunting and meat exchange reveals the hidden importance of these activities in the traditional social reciprocity systems that provide the essential "glue" to bind together dispersed households and communities. In the current climate with its strong emphasis on markets and economic incentives, the paper draws attention to the overlooked importance of kinship relations, and how collaboration among kin and neighbors is enacted through hunting and meat distribution, important bonds which heavily influence both the physical and cultural landscape, and that could be undermined by market relationships.

Combining both social and agricultural science perspectives, Brazilian scholars Noemi Porro, Iran Veiga, and Dalva Mota explore the invisible social underpinnings of Amazonian livelihoods and identities in their paper on the emergence and risks of new political identities and alliances among women who traditionally earn their living from cracking open and selling the nuts from the babassu palm. The paper traces how a large population with diverse origins and land tenure situations was able to unite around common values and demands based on their traditional right to babassu as a common-use resource, an integral complement to their agriculturally-based livelihood in the contested territories of eastern Amazonia, where babassu forests are widespread. The babassu breakers' social movement has faced challenges of clientelism, imposition of outside agendas, and competing agendas for mobilization, which they are addressing through a dynamic process of social learning to seek double legitimacy: with members of communities that participate in their social movement, and with a network of external allies.

The last two papers in the volume explore transboundary landscapes in the Amazon region, at two distinct scales: regional and continental. Based on fieldwork in the transboundary borderlands between Peru and Brazil in southwestern Amazonia, the paper by geographer David Salisbury with Peruvian colleagues José Borgo López and Jorge Vela Alvarado explores how the Asháninka indigenous peoples on both sides of the border have responded to resource wars that historically have driven multi-scalar, dynamic local changes in resource boundaries. With a common culture and history, the Asháninka defy notions of the "empty" Amazon by actively engaging in the defense of their borderland territories, where biogeographical boundaries delimiting resources and ecosystems blend in complex ways with national political boundaries straddled by these indigenous people. In the transboundary political ecology, the relative success of these groups depends on their ability to negotiate land rights and other forms of recognition by the state.

The paper by geographer Sonja Pieck analyzes the Initiative for the Integration of Regional Infrastructure in South America (IIRSA), an ambitious multilateral program designed to reshape the entire landscape of the continent through 500 separate infrastructure projects, and the responses of transnational activist networks to these proposals as they seek to engage with states in a negotiation process. The paper explores the exercise of citizenship in a neoliberal setting, on the part of coalitions of activists who seek to institutionalize civil society participation through demands for greater transparency. Going beyond postdevelopment resistance to modernizing projects, these groups are negotiating the terms of their engagement with the state.

In her concluding paper for this volume, geographer Susanna Hecht comments on the emergence of a grassroots "Amazon Nation," comprised of diverse local peoples, identities, and places, constructed through the assertion of new forms of citizenship, identity, and socio-environmentalisms as part of a new "statecraft" from below. She traces the roots of these new Amazonian geographies to the importance of Amazonia in the construction of the Brazilian modernist state, and the multiple forms of resistance that arose to contest deforestation and commodity expansion at the expense of culturally-rooted local popula-tions linked to Amazonian territories. As Brazil steps to the forefront of global markets and politics, the emerging voices of the Amazon Nation will continue to demand their place in the debate, and to defend their uniquely Amazonian identities, livelihoods and landscapes.

Conclusion

The papers in this volume present vivid dimensions of local Amazonian peoples as proactive, place-based social and political actors involved in a dialogue with current dominant development models, and proposing alternatives inspired by their own social and historical experiences in concrete Amazonian territories. Despite their historical invisibility, these are "living images of social actors raising their voices to speak about specific public policies" (Porro *et al.* 2011). The diversity of specific groups— transnational activist networks; borderland indigenous peoples; rubber tappers; urban dwellers; riverside and forest inhabitants; palm nut breakers; primordial and newly recreated indigenous groups—reflects the distinct histories of migration, territorial conflict, resource exploitation and culture conflict across the communities, regions and populations of the Amazon basin (Little 2001). Their social movements have achieved various degrees of empowerment through organizing and alliances, as well as shifting their livelihoods and their discourses to construct new identities.

Alongside well-funded national and multinational development efforts to extend roads and dams throughout the Amazon region, local people and their allies also are involved in surprisingly successful processes of grassroots transformative social change, as exemplified by the policy successes of the rubber tappers social movement in creating and expanding new land reform concepts. These social movements, operating across diverse scales through networks and alliances, assert the rights of citizens to question the social and environmental impacts of large-scale development projects, and to propose alternatives based not only on economic goals, but also on such "moral economy" principles as justice, autonomy, common property, reciprocity, and mutual care. These movements and their proposals have resonated with emerging national and international environmental alliances and proposals over the past two decades, producing new and complex partnerships among diverse actors seeking to resist and engage state policies, while articulating alternative discourses and policies more appropriate to their local contexts.

Acknowledgements

The papers in this volume grew out of a session organized by Jacqueline Vadjunec and Alyson Greiner on Amazonian Geographies held at the Annual Meeting of the Association of American Geographers (AAG) in Washington, D.C. in April 2010, and a three-panel symposium organized by Marianne Schmink and Jacqueline Vadjunec on Changing Identities, Landscapes, Livelihoods and Discourses in Brazilian Amazonia, held at the Conference of the Society for Amazonian and Andean Studies, University of Florida, Gainesville, Florida November 2010. We would like to thank both the presenters and the audiences for their participation and insights, as well as the Cultural and Political Ecology Specialty Group and Cultural Geography Specialty Group for session sponsorship at the AAG. We are grateful to the many anonymous reviewers, whose comments much improved the contents in this volume. The editors also thank Alyson Greiner, the Editor of the *Journal of Cultural Geography*, for her patience, persistence, and many helpful comments and assistance throughout the entire editing process. Lastly, we thank Michael P. Larson for his cartographic expertise and willingness to help out in the eleventh hour, and Jay H. Jump for his keen eyes and helpful editorial assistance.

References

Achard, F., *et al.*, 2002. Determination of deforestation rates of the world's humid tropical forests. *Science*, 297, 999–1002.

Allegretti, M.H., 1990. Extractive reserves: an alternative for reconciling development and environmental conservation in Amazonia. *In*: A.B. Anderson, ed. *Alternatives to deforestation: steps toward sustainable use of the Amazon rain forest*. New York: Columbia University Press, 252–264.

Almeida, A.L.O. and Campari, J.S., 1995. *Sustainable settlement in the Brazilian Amazon*. New York: Oxford University Press.

Almeida, A.W.B., 2008. *Terras de quilombos, terras indígenas, babaçuais livres, castanhais do povo, faxinais e fundos de pasto: terras tradicionalmente ocupadas* 2nd ed. Manaus: Federal University of Amazonas, Projeto Nova Cartografia Social da Amazônia, Coleção "Tradição e ordenamento jurídico," vol. 2.

Alvarez, S., Danino, E. and Escobar, A., eds., 1998. *Cultures of politics, politics of cultures: re-visioning Latin American social movements*. Boulder: Westview Press.

Balée, W., 1994. *Footprints of the forest: Ka'apor ethnobotany—the historical ecology of plant utilization by an Amazonian people*. New York: Columbia University Press.

Black, R., 1990. Regional political ecology in theory and practice: A case study from northern Portugal. *Transactions of the Institute of British Geographers*, 15, 35–47.

Blaikie, P. and Brookfield, H.C., 1987. *Land degradation and society*. London: Methuen.

Bebbington, A.J., 2001. Globalized Andes? Livelihoods, landscapes and development. *Ecumene*, 8 (4), 414–436.

Bebbington, A.J. and Batterbury, S., 2001. Transnational livelihood and landscapes: political ecologies of globalization. *Ecumene*, 8 (4), 369–380.

Bolaños, O., forthcoming 2011. Redefining identities, redefining landscapes: indigenous identity and land rights struggles in the Brazilian Amazon. *Journal of Cultural Geography*.

Browder, J.D., 1994. Surviving in Rondonia: the dynamics of colonist farming strategies in Brazil's northwest frontier. *Studies in Comparative International Development*, 29 (3), 45–69.

Browder, J.D. and Godfrey, B.J., 1997. *Rainforest cities: urbanization, development, and globalization of the Brazilian Amazon*. New York: Columbia University Press.

Bryant, R. and Bailey, S., 1997. *Third World Political Ecology*. London: Routledge.

COICA (Coordinadora de las Organizaciones Indigenas de la Cuenca Amazónica), 1996. *Amazonia: economia indigena y mercado. Los desafios del desarollo autónomo*. Quito: COICA/Oxfam America.

Colchester, M., 2000. Self-determination or environmental determinism for indigenous peoples in tropical forest conservation. *Conservation Biology*, 14 (5), 1365–1367.

Conklin, B.A. and Graham, L.R., 1995. The shifting middle ground: Amazonian Indians and eco-politics. *American Anthropologist*, 24 (4), 711–737.

Cronon, W., 1994. Cutting loose or running aground? *Journal of Historical Geography*, 20 (1), 38–43.

Cunha, M.C. and Almeida., M.W.B., 2000. Indigenous people, traditional people and conservation in the Amazon. *Journal of the American Academy of Arts and Sciences*, 129 (2), 315–338.

Davidson, J., Bondi, L. and Smith, M., eds., 2005. *Emotional geographies*. Burlington, VT: Ashgate Press.

Dean, W., 2002. *Brazil and the struggle for rubber*. New York: Cambridge University Press.

Denevan, W.M., 1996. A bluff model of riverine settlement in prehistoric Amazonia. *Annals of the Association of American Geographers*, 86 (4), 654–681.

Duncan, J.S. and Duncan, N.G., 2004. *The politics of the aesthetic in an American suburb*. New York: Routledge.

Escobar, A. and Paulson, S., 2005. The emergence of collective ethnic identities and alternative political ecologies in the Colombian Pacific rainforest. *In*: S. Paulson, ed. *Political ecology across spaces, scales, and social groups*. New Brunswick: Rutgers University Press, 257–278.

Fearnside, P.M., 1997. Limiting factors for development of agriculture and ranching in Brazilian Amazonia. *Revista Brasileira de Biologia*, 57, 531–549.

Fisher, W.H., 1994. Megadevelopment, environmentalism, and resistance: the institutional context of Kayapó indigenous politics in central Brazil. *Human Organization*, 53 (3), 220–232.

Freire, P., 1970. *Pedagogy of the oppressed*. New York: Herder & Herder.

Glasner, B. and Woods, W.I., eds., 2004. *Amazonian dark earths: explorations in space and time*. New York: Springer.

Goulding, M., Barthem, R., and Ferreira, E.J.G., 2003. *The Smithsonian atlas of the Amazon*. Washington, DC: Smithsonian Books.

Guimâraes, R., 1991. *Ecopolitics of development in the Third World*. Boulder: L. Reinner Pub.

Hall, A., 1997. *Sustaining Amazonia: grassroots action for productive conservation*. Manchester: Manchester University Press.

Harris, M., 2000. *Life on the Amazon: the anthropology of a Brazilian peasant village*. Oxford: Oxford University Press.

Harris, M., 2007. Introduction: ways of knowing. *In*: M. Harris, ed. Ways of knowing: anthropological approaches to crafting experience and knowledge. Oxford: Berghahn Books, 1–26.

Hecht, S., 1993. The logic of livestock and deforestation in Amazonia. *Bioscience*, 43 (1), 687–695.

Hecht, S., 2004. The last unfinished page of Genesis: Euclides da Cunha and the Amazon. *Historical Geography*, 32, 43–69.

Hecht, S., forthcoming 2011. *The scramble for the Amazon: imperial contests and the tropical odyssey of Euclides da Cunha*. Chicago: University of Chicago Press.

Hecht, S. and Cockburn, A., 1990. *Fate of the forest: developers, destroyers and defenders of the Amazon*. Verso: Routledge.

Heckenberger, M.J., *et al.*, 2007. The legacy of cultural landscapes in the Brazilian Amazon: implications for biodiversity. *Philosophical Transactions of the Royal Society*, 362, 197–208.

Heckenberger, M.J., *et al.*, 2008. Pre-colombian urbanism, anthropogenic landscapes, and the future of the Amazon. *Science*, 321 (5893), 1214–1217.

Heynen, N., Kaika, M., and Swyngedouw, E., 2006. Urban political ecology: politicizing the production of urban natures. *In*: N. Heynen, M. Kaika, and E. Swyngedouw, eds. *The nature of cities: urban political ecology and the politics of urban metabolism*. Oxford: Routledge, 1–20.

Hugill, P.J. and Foote, K.E., 1994. Re-reading cultural geography. *In*: K.E. Foote, *et al.*, eds. *Re-reading cultural geography*. Austin: University of Texas Press, 9–23.

Jackson, J. and Warren, K., 2005. Indigenous movements in Latin America, 1992–2004: controversies, ironies, new directions. *Annual Review of Anthropology*, 34, 549–573.

Jackson, P., 1989. *Maps of meaning: an introduction to cultural geography.* London: Unwin Hyman.

Keck, M., 1995. Social equity and environmental politics in Brazil: lessons from the rubber tappers of Acre. *Comparative Politics*, 27 (4), 409–424.

Keith, M. and Pile, S., 1993. Introduction part 1: the politics of place. *In*: M. Keith and S. Pile, eds. *Place and the politics of identity.* London: Routledge, 1–21.

Langer, E.D. and Muñoz, E., eds., 2003. *Contemporary indigenous movements in Latin America.* Wilmington: Scholarly Resources.

Ley, D., 2001. Introduction: landscapes of dominance and affection. *In*: P.C. Adams, S. Hoelscher, and K.E. Till, eds. *Textures of place: exploring humanist geographies.* Minneapolis: University of Minnesota Press, 3–7.

Little, P.E., 2001. *Amazonia: territorial struggles on perennial frontiers.* Baltimore: Johns Hopkins University Press.

London, M. and Kelly, B., 2007. *The last forest: the Amazon in the age of globalization.* New York: Random House.

Lowenthal, D., 2001. Making a pet of nature. *In*: P.C. Adams, S. Hoelscher, and K.E. Till, eds. *Textures of place: exploring humanist geographies.* Minneapolis: University of Minnesota Press, 84–92.

Ludewigs, T., 2006. *Land-use decision making, uncertainty and effectiveness of land reform in Acre, Brazilian Amazon.* Dissertation (PhD). Indiana University.

Malpas, J.E., 2007. *Place and experience: a philosophical topography.* Cambridge: Cambridge University Press.

Meggers, B.J., 1971. *Amazonia: man and culture in a counterfeit paradise.* Chicago and New York: Aldine.

Miller, V., Hallstein, M., and Quass, S., 1996. Feminist politics and environmental justice: women's community activism in West Harlem, New York. *In*: D. Rocheleau, B. Thomas-Slayter, and E. Wangari, eds. *Feminist Political Ecology: Global Issues and Local Experiences.* London: Routledge, 62–85.

Minzenberg, E. and Wallace, R., forthcoming 2011. Amazonian agriculturalists bound by subsistence hunting. *Journal of Cultural Geography.*

Moran, E., 1981. *Developing the Amazon.* Bloomington: Indiana University.

Nash, C., 2000. Performativity in practice: some recent work in cultural geography. *Progress in Human Geography*, 24 (4), 653–664.

Nepstad, D.C., *et al.*, 2006. Conservation areas stop forest clearing. *Conservation Biology*, 20, 65–73.

Nova cartografia social da Amazônia [online]. Available from: http://www.novacartografiasocial.com [Accessed 3 December, 2010].

Nugent, S., 1993. *Amazonian Caboclo society: an essay on invisibility and peasant economy.* Oxford: Berg Publishers.

Olwig, K.R., 2001. Landscape as a contested topos of place, community, and self. *In*: P.C. Adams, S. Hoelscher, and K.E. Till, eds. *Textures of place: exploring humanist geographies.* Minneapolis: University of Minnesota Press, 93–117.

Parker, E., ed., 1985. *The Amazon caboclo: historical and contemporary perspectives.* Williamsburg, VA: College of William and Mary.

Peet, R. and Watts, M., 1996. *Liberation Ecologies.* London: Routledge.

Perreault, T., 2001. Developing identities: indigenous mobilization, rural livelihoods, and resource access in Ecuadorian Amazonia. *Ecumene*, 8 (4), 381–413.

Perz, S.G., Aramburú, C., and Bremner, J., 2005. Population, land use and deforestation in the Pan Amazon Basin: a comparison of Brazil, Bolivia, Colombia, Ecuador, Perú and Venezuela. *Environment, Development and Sustainability*, 7 (1), 23–49.

Pieck, S.K., forthcoming 2011. Beyond postdevelopment: civic responses to regional integration in the Amazon. *Journal of Cultural Geography*.

Pieck, S.K. and Moog, S.A., 2009. Competing entanglements in the struggle to save the Amazon: the shifting terrain of transnational civil society. *Political Geography*, 28 (7), 416–425.

Porro, N., Veiga, I., and Mota, D., forthcoming 2011. Traditional communities in the Brazilian Amazon and the emergence of new political identities: the struggle of the quebradeiras de coco babaçu—babassu breaker women. *Journal of Cultural Geography*.

Posey, D.A., 2001. Alternatives to forest destruction: lessons from the Mêbêngôkre Indians. *In*: S.E. Place, ed. *Tropical rainforests: Latin American nature and society in transition*. Wilmington: Scholarly Resources Incorporated, 179–186.

Povos Indígenas no Brasil. Introdução [online]. Available from: http://pib. socioambiental.org/pt/c/no-brasil-atual/quem-sao/introducao [Accessed 3 December, 2010].

Ramos, A.R., 1998. *Indigenism: ethnic politics in Brazil*. Madison: University Wisconsin Press.

Revkin, A., 2004. *The burning season: the murder of Chico Mendes and the fight for the Amazon rain forest*. New York: First Island Press.

Roach, J., 2003. Amazon tribes: isolated by choice? National Geographic [online]. Available from: http://news.nationalgeographic.com/news/2003/03/0310_030310_invisible1.html [Accessed 1 December 2010].

Robbins, P., 2004. *Political ecology: a critical introduction*. Oxford: Blackwell.

Rocheleau, D., Thomas-Slayter, B., and Wangari, E., 1996. Gender and environment: a feminist political ecology perspective. *In*: D. Rocheleau, B. Thomas-Slayter, and E. Wangari, eds. *Feminist political ecology: global issues and local experiences*. London: Routledge, 3–22.

Roux, J., 2001. From the limits of the frontier: or the misunderstandings of Amazonian geo-politics. *Revista De Indias*, 61 (223), 513–39.

Salisbury, D.S., Borgo López, J., and Vela Alvarado, J.W., forthcoming 2011. Transboundary political ecology in Amazonia: history, culture, and conflicts of the borderland Asháninka. *Journal of Cultural Geography*.

Salisbury, D. and Schmink, M., 2007. Cows versus rubber: changing livelihoods among Amazonian extractivists. *Geoforum*, 38 (6), 1233–1249.

Schmink, M. and Cordeiro, M.L., 2008. *Rio Branco: a cidade da Florestania*. Bélem: Universidade Federal do Parà.

Schmink, M. and Wood, C.H., 1992. *Contested frontiers in Amazonia*. New York: Columbia University Press.

Scott, J., 1976. *The moral economy of the peasant*. New Haven, CT: Yale University Press.

Skole, D.L., Chomentoski, W.H., Salas, W.A., and Nobre, A.D., 1994. Physical and human dimensions of deforestation in Amazonia. *BioScience*, 44, 314–322.

Slater, C., 2002. *Entangled Edens: visions of the Amazon*. Berkeley: University of California Press.

Smith, N., 1982. *Rainforest corridors: the transamazon colonization scheme*. Berkeley: University of California Press.

Smith, N., 1999. *The Amazon river forest: a natural history of plants, animals, and people*. New York: Oxford University Press.

Stokes, C.E., 2000. *The Amazon bubble: world rubber monopoly*. Fort McKavett: Charles Stokes.

Treece, D., 1994. The militarization and industrialization of Amazonia: The Calha Norte and Grande Carajas Programs. *In*: S.E. Place, ed. *Tropical rainforests: Latin American nature and society in transition*. Wilmington: Scholarly Resources Incorporated, 62–70.

Tuan, Y-F., 1974. *Topophilia: a study of environmental perception, attitudes, and values*. New York: Columbia University Press.

Turner II, B.L., 1997. Spirals, bridges, and tunnels: engaging human-environment perspectives in geography. *Ecumene*, 4 (2), 196–217.

Turner II, B.L. and Robbins, P., 2008. Land-change science and political ecology: similarities, differences, and implications for sustainability science. *Annual Review of Environment and Resources*, 33, 295–316.

Vadjunec, J.M., Schmink, M., and Gomes, C.V.A., forthcoming 2011. Rubber tappper citizens: emerging places, policies, and shifting identities in Acre, Brazil. *Journal of Cultural Geography*.

Walker, R., Moran, E., and Anselin, L., 2000. Deforestation and cattle ranching in the Brazilian Amazon: external capital and household processes. *World Development*, 28 (4), 683–699.

Zimmerer, K., 2006. Cultural ecology: at the interface with political ecology—the new geographies of environmental conservation and globalization. *Progress in Human Geography*, 30 (1), 63–78.

Waiwai fractality and the arboreal bias of PES schemes in Guyana: what to make of the multiplicity of Amazonian cosmographies?

Laura H. Mentore

Department of Sociology and Anthropology, University of Mary Washington, Fredericksburg, VA, USA

This paper concerns the often-assumed universal translatability of the environment as it exists in conservation discourses and policies in Amazonia. It focuses on Guyana's Low Carbon Development Strategy (LCDS), a government-led strategy to achieve socioeconomic development goals through foreign capital investment and trading in rights over "eco-services" provided by the country's forests. Guyana's indigenous peoples have been identified as one of the largest stakeholder groups for the LCDS and as such, are particular targets for translating it. Drawing from my research with indigenous Waiwai, I explore some cosmographic and cosmological differences between the arboreal, unilinear, and commodified environment portrayed in the LCDS and the fractal, recursive, and mediated environment portrayed in Waiwai discourse and practice. The way in which I understand the Waiwai lived environment is informed by their very ways of moving through it: along meandering and indirect interpretations shaped by social memories, nuanced impressions, physical and affective experiences. As the role of ethnographer is increasingly defined in terms of translation, there may be grounds for considering the differences between the knowledge-politics of direct, mechanical translation practices often found in consultations with indigenous communities and the more long-term, roundabout interpretive processes of ethnographic fieldwork and writing.

One of the main problems raised by the editors of this collection of papers is the enduring tendency to think of Amazonia as a "homogeneous place" despite the abundance of historical, socio-geographic and cultural evidence to the contrary. In the interest of further problematizing this tendency, this paper focuses on a cosmographic interface between distinct environments, and considers the knowledge-politics of "interpreting" versus "translating" them. It is these complex interfaces between environments which, for the sake of scholarly discussion and debate, we collectively refer to as Amazonia.

Just as it has been wrongly assumed that the ecology of Amazonia is a homogeneous and pristine system devoid of human modifications, so too is it increasingly taken for granted that a single knowledge system or epistemology can accurately and holistically capture its meaning and value. While cultural ecology and ethnobotany have taken significant steps toward dismantling the former (e.g., Balée 1994; Descola 1996; Posey 2007), both assumptions often are integral to the manner in which indigenous communities in Guyana are consulted by governmental and non-governmental agencies seeking their "informed consent" for conservation initiatives. Convinced of the seamless correspondence between the environment and their representations of it, the only intellectual work that remains to be done in consultations from the perspective of the NGO or government agency is essentially of a translational nature (Errington and Gewertz 2001; Zerner 2003; West 2005). One of the more dubious phrases used to describe such work is "capacity building." Various projects, exercises, technologies, models and diagrams are introduced to indigenous communities with the intention of "building up" their "capacity" to know and relate to the environment according to their representations and translations of it. Yet as Maranhão writes:

> The pure and simple notion of translation as a transportation of meaning from one semiotic system to another is always embedded in a history defining the political relations between two systems ... There are always power-related reasons justifying why something 'needs' to be spoken or written in another language. (2003, p. 64)

West (2005, 2006), an anthropologist working with the Gimi in the Eastern Highlands of Papua New Guinea, argues that conservation and development projects often facilitate a "generification of culture" (West 2005, p. 633, from Errington and Gewertz 2001) through translation practices that assume local knowledge can be accounted for in conservation models of the environment as "resources to be used" and "knowledge to be acted upon" (West 2005, p. 633; Ellis and West 2004). Political ecologists and environmental anthropologists are increasingly involved in this process as a result of expectations that they can "translate local knowledge into categories that scientists can easily understand and assimilate into their epistemologies" (West 2005, p. 632). Such attempts are invariably motivated by the best of intentions, given that "scientists who do not see the value of local knowledge are much more likely to suggest conservation polices that are not socially equitable" (West 2005, pp. 632–633; Harper 2002). Yet West, and others such as Murray Li

(2007) raise important questions about the politics of knowledge and translation in conservation and development projects, particularly concerning how to situate research based on sustained ethnographic fieldwork and writing in relation to such politics.[1] Murray Li (2007, pp. 3–4) reflects on her self-described inability to "translate" her years of experience in researching rural life in Indonesia into the "world of projects" inhabited by development administrators, referring to it as a "predicament" that enables her to more critically reflect upon "what ways of thinking, what practices and assumptions are required to translate messy conjectures, with all the processes that run through them, into linear narratives of problems, interventions and beneficial results".

With such issues in mind, this paper critically examines Payment for Environmental Service (PES) schemes currently emerging in Guyana and their claim of universal translatability to all interested parties and actors—from overseas investors to government representatives and indigenous communities. Like all environments, the environment as defined in PES schemes exists only within the framework of ideas, statements, practices and relations of power that sustain and reproduce it as reality or "truth." To stress this point, aspects of the PES environment are considered in a comparative light, alongside glimpses of the Waiwai lived environment. I focus on what I interpret to be the arboreal, unilinear, and commodity-based environment that emerges through PES discourses and practices, and the fractal, recursive, and mediated environment that emerges through Waiwai ways of knowing and being. The apparent differences between the two are significant given that the Waiwai, as an indigenous "stakeholder group," are principal targets for the translation of PES schemes in Guyana.

This paper is based on 27 months of anthropological field research in Guyana, carried out in various periods between 2004 and 2010 (including doctoral research from May 2005 to March 2006), primarily in the Waiwai communities of Erepoimo (which continues to be officially registered as Parabara) and Masakenyaru. Shorter stays in the Wapishana communities of Karaudarnaua and Aishalton (approximately 30 and 60 miles north of Erepoimo, respectively), the Makushi community of Surama (in the Rupununi savannas of central Guyana), and the capital city of Georgetown have also greatly contributed to my understanding of issues discussed here (Figure 1). The research entailed long-term participant observation and informal and formal interviews with residents of Waiwai and Wapishana communities, as well as observations of community consultations and interviews with NGO and government representatives. Newspaper articles, websites, and publicly distributed literature in Guyana were also important secondary sources of information on the Low Carbon Development Strategy. Out of confidentiality, the names of interlocutors have either been changed or omitted.

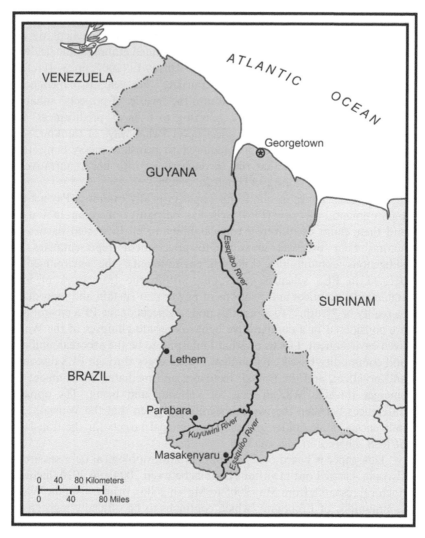

Figure 1. Map of Guyana indicating field sites of Erepoimo, Masakenyaru Lethem, and Georgetown.

The limits of translation are approached here by considering a politics of knowledge that goes beyond surface or representational differences to constitute a "clash of cosmographies" (Rubenstein 2004, p. 135; see also Little 2001). My use of the term cosmography is informed by Little (2001, p. 5), as encompassing "the symbolic and affective relationship a group maintains with its biophysical environment, which creates bonds of identity between a social group and a geographical area . . .". This is underpinned by the understanding that all environments are socially imagined and produced spaces (Little 2001; Lefebvre 1991; Smith 1990). My understanding of the interrelationship between theory and ethnography is further inspired by Yi-Fu Tuan's (1977, p. 6) foundational contributions to cultural geography, that although lived experience often "resists ready communication," any account of cultural forms of space and place nonetheless must attempt to understand "the different modes of experience (sensorimotor, tactile, visual, conceptual)" through which people know and construct reality. It also can be read as a preliminary attempt at responding to his call for more humanistic interpretations of space and place, as "images of complex—often ambivalent—feelings" (Tuan 1977, p. 7).

Before going further, a brief example may help to illustrate the kind of translation assumptions I have in mind. In 2004, a team of Conservation International-Guyana (CIG) staff members and government representatives held consultations in several Waiwai and Wapishana communities regarding the establishment of a protected area in the region. A man from one of the communities was employed as the "Waiwai translator." The difficulty of the task before him became apparent no sooner than he was faced with the term "protected area," which of course was spoken constantly by the consultation team. At one point, he tried explaining to the team that he was struggling to explain "protected area" in the Waiwai language and was at a loss for words. They simply urged him to come up with the closest possible translation, and the meetings continued.

In his attempt to translate a concept which he essentially regarded as beyond translation, he turned to *wokpo*—the Waiwai term for lines of fallen timbers that are sometimes arranged between adjacent manioc gardens of different household families. By way of analogy, he tried to provide his audience with an image of protected area through reference to this most familiar kind of line. In the end, most communities did not reach a consensus on the protected area and the project was abandoned with the exception of a protected area around Masakenyaru village at the upper Essequibo River. Yet in their final report, the consultation team cites such acts of translation as examples of their compliance with the ethical protocols of the NGO, and the fact that these communities did not reach a consensus ultimately was attributed to internal socio-political frictions rather than anything to do with the consultation process itself.

Six years after these consultations, the same man—one of my closest friends and interlocutors—has continued to talk with me about the many conceptual differences between protected area and *wokpo*, expressing both frustration at being pressured to provide a translation and concern that the lack of a "better translation" may have been a factor in his community not supporting the initiative. In my ongoing efforts to understand and interpret Waiwai notions of space and environment, his reflections upon this "failed analogy" have been highly insightful, if not moving. *Wokpo* lines are probably the closest visual correlate in the material environment of the Waiwai to the kinds of lines and bounded spaces depicted on CIG's maps of projected protected areas, and one can understand why the analogy would come to mind. Yet the social purpose of *wokpo* differs dramatically from the boundary lines of protected areas. The manioc gardens encompassed by the former could be described as explicitly "unprotected" areas in the sense that their contents are broadly distributed and consumed, they are considered openly available to people travelling through the area, and women frequently bring guests to their gardens to fill their baskets with produce. Forest animals also freely enter the gardens at dawn and dusk—sometimes to the devastation of crops— and no attempt is made to keep them out. The *wokpo* themselves decompose and vanish from sight as gardens are left fallow for a decade or longer. *Wokpo* signify lines of reciprocity across households (and across human and non-human spaces), not boundaries of exclusion between them. My point is not that a more accurate translation could have been arrived at if CIG had been more patient or thorough, but rather, that such a translation may not exist. This is where I suggest that for analytic purposes, it may be worth noting a difference between translation practices and interpretative processes with regard to the politics of knowledge in conservation-development schemes.

My working concept of interpretation can be likened to what Maranhão (2003, p. 80) describes as "an anthropology against translation," or the interminable practice of dwelling in the dynamic difference between word and thing. What I have in mind are the attempts we make in ethnographic fieldwork and writing to partially elucidate the meaning of relational forms and processes through a long-term commitment to meandering excursions along and around multiple interpretive possibilities. Above all, Maranhão explains, such an approach requires patience with the immediate meaningless of the said, privileging of the saying over and above the said, and embracing of the view that "the interesting meaning of what it studies lies in the surplus of signification ..." (2003, p. 82). Schuler Zea (2009) adopts a similar stance in her recent essay on the Waiwai in Brazil, where she likens her meanderings in the space of ethnographic writing to Waiwai ways of speaking and moving through the environment along *yesamari* or "detours." As she writes, "...it is the process and not the goal that appears with an accentuated meaning—not as a means to a previewed end, but in the form of a persistent detour that figures as reference for the interminable *identité-à-faire* Waiwai" (Schuler Zea, p. 211).[2]

PES and Guyana's vision of development through the carbon market

The concept of PES (Payment for Environmental Services) represents a relatively new component of green trading—financial trading in "eco-friendly" products, services and technologies thought to yield capital gains for the trader, positive outcomes for the environment and oftentimes, positive outcomes for "developing," rural, and/or indigenous populations in ecologically valuable but threatened environments. Within this framework, PES schemes are premised on the specific notions that certain ecosystems provide valuable "eco-services" such as carbon sequestration and rainfall generation, which help to mitigate the negative effects of global climate change; that the value of these services can and should be calculated in monetary terms; and that persons who act as fiduciaries of such eco-services should be monetarily compensated for the cost of their provision (e.g., losses incurred by not clearing the forests for commercial logging, agriculture, mining, etc.). PES schemes are commonly described as correcting a market failure—for example, the failure of markets to recognize the greater value of trees alive rather than dead (Jagdeo 2008, p. 1).

Although market-based approaches to environmental protection have existed for several decades, PES schemes mark an important shift from "preventing negative environmental externalities through eco-taxes and charges based on the polluter-pays principle ... to creating positive environmental externalities ... through appropriate economic incentives, often in the form of subsidies or other environmental programs ..." (Maynard and Paquin 2004, p. 1). This shift has received strong support from developing countries such as Guyana, which have widely accused the previous carbon credit schemes of benefitting wealthier, more industrialized countries with higher levels of carbon emissions. The Guyanese government has recognized the distinct prospect of benefiting from compensatory PES schemes, given that over 80% of the country's landmass, roughly 15 million hectares or 58,000 square miles, (an area larger than England) is standing forest with unparalleled levels of biodiversity, low human population density, and numerous endangered species of plants and animals. In June 2009, the government launched the first draft of its nationwide Low Carbon Development Strategy—six months in advance of the United Nations' Climate Change Conference in Copenhagen (Office of the President, Government of Guyana 2010). President Jagdeo has proposed the conservation of *all* of Guyana's standing forests if foreign governments and venture capitalists will invest in the "global eco-services" they provide, with an initial emphasis on carbon sequestration. He has also proposed to direct revenue from the sale of rights over these eco-services toward education and healthcare reform, as well as much-needed repairs to the eroding seawalls and drainage system along the more densely populated coast of Guyana. The PES schemes that will propel the LCDS are likely to be coordinated initially through the United Nations program REDD+ (Reducing Emissions Through Deforestation and Forest Degradation), a climate change mechanism for paying countries to conserve their forests, which is currently laying a foundation for future market-based conservation across the Guianas.

It is significant to note that the indigenous peoples of Guyana (who comprise an estimated 9% of Guyana's population) hold legal title to 14% of the country's territory, much of which is forested (Sukhai 2010). The 300 Waiwai at Masakenyaru village alone have collective title to approximately 6.2 million hectares (2,400 square miles) of almost entirely forested land, including the headwaters of Guyana's largest river, the Essequibo, and one of the country's least protected border zones with Brazil and Suriname. Thus, perhaps it is not surprising that when I was in Guyana in 2009 and 2010, the government was busily conducting consultations in indigenous communities throughout the interior. From what I have observed, these consultations largely amount to an intensive effort to translate the LCDS and its notions of carbon storage, climate regulation and sustainable development through conservation as efficiently as possible, with recourse to diagrams, maps and models as well as speeches by politicians known for their support among indigenous peoples. Translators are employed for some consultations, and it appears that efforts have been underway since 2009 to produce written translations of the LCDS into five indigenous languages—an initiative that has sparked considerable debate (Kaiteur News 2010). In consultations, it is stressed that the preliminary PES schemes would pertain to state lands only, and that indigenous communities would then have the ability to "opt in" or "opt out" of PES schemes pertaining to their lands at any time (Government Information Agency of Guyana 2010, pp. 3–4). It is also stressed, however, that if they should choose to "opt in," this would generate significant revenue for education, healthcare and other sustainable development projects in their communities.

Having outlined some basic components of the LCDS, we can consider what I refer to as its "arboreal bias." In Velásquez Runk's insightful study of the arboreal bias in conservation initiatives in the indigenous Wounaan territory in Eastern Panama, she describes the state's approach as aimed at "taming the region ... and over-simplifying its heterogeneous landscape through arborescent models in which complexities are represented through binary logics ... such as nature-culture or forested-deforested" (2009, pp. 461–462). She draws from Deleuze and Guattari's (2007) notion of arborescent knowledge—the image of the root-tree that graphically depicts hierarchical systems of order and relationality, where knowledge is constantly re-centered and confined to the binary logic of dichotomies and linear bifurcations, a regime of thought which "has never reached an understanding of multiplicity" and instead, "plots a point, fixes an order ... and proceeds by dichotomy" (Deleuze and Guattari 2007, pp. 5–7).

Far from new or novel, the arboreal bias in conservation and PES schemes can be traced to a long-standing Euro-American tradition of using tree imagery to depict upward, unilinear forms of movement or development from past to future—from the emergence of modernity and urbanity out of a "forested antecedence" (Harrison 1992, p. 1) to the popular use of the "family tree" to record and display descent-based kinship ties and the infamous use of tree imagery by early evolutionary anthropologists such as Spencer to "spatialize time" according to "classificatory schemes of subsumption and hierarchy" (Fabian 1983, p. 15). My interest lies in the much related symbolic and economic role of trees in contemporary development discourses in Guyana. There too, trees are playing a preeminent role in the construction of a new "global ecological imaginary" (Conklin and Graham 1995) wherein powerful correlations are being drawn between the capacity of trees to absorb and store carbon, the socio-economic development of the nation, and the mitigation of global climate change.

As Hennessy (1978, p. 12) has discussed, the plethora of frontiers that have continually opened, closed and re-opened throughout Amazonia's diverse histories often are marked by "cyclical booms in different commodities". Little (2001, pp. 2–3) argues that throughout these histories "each wave was based in the new desires, knowledge systems, technologies, and forms of social organization brought by social actors into Amazonia and was marked by the resources they extracted, the markets where they traded, and the biophysical effects they produced". In this sense, the arboreal bias in PES schemes such as the LCDS can be seen as the central ideological component of yet another regional frontier in Amazonia. Unlike previous commodity booms also based on trees, such as the rubber boom, the commodity value of trees in PES schemes is driven not by the physical extraction of their material properties but rather, investment and trading in what I would describe as the intangible services they are thought to provide when left standing.

As another example of forest conservation being treated as synonymous with environmental conservation (West 2005, p. 462), the arboreal bias of the LCDS is of much political and economic consequence for indigenous communities located in non-forest environments such as tropical savanna—notably among the Makushi and Wapishana in the Rupununi savannas. A number of communities in the Rupununi are currently disputing their land titles, which in most cases are based on surveys conducted during or immediately after British colonial occupation of Guyana in the 1960s. These surveys often failed to take into account the forested areas beyond village parameters, which communities extensively rely upon for farmland, hunting grounds, housing materials, medicinal plants, and other purposes (L. Mentore 2010). Through what Hennessy (2003) refers to as the "villagization" of indigenous lands, indigenous peoples whose lifeways and environments reflect patterns of socio-spatial mobility are increasingly tied (economically, politically and cartographically) to fixed "places-as-points" (Ingold 2007). Although steps have been taken in recent years to survey or re-survey lands, and many communities have received land title or extensions to pre-existing titles, the arboreal bias in PES schemes nonetheless raises serious concerns for communities still seeking title to forested portions of their socio-geographic territory. For these reasons, it stands to compromise indigenous support of the LCDS and could spur tensions between communities based on whether their titled lands include forested areas or only savanna—tensions already set in motion by previous conservation initiatives premised on the same arboreal bias.

In keeping with Western perceptions of environments as knowledges to be acted upon according to their potential commodity value (Ellis and West 2004), the driving question for the LCDS has become how to determine a commodity value of standing trees that surpasses that of lumber and/or commercial agricultural land. As Rubenstein (2004, p. 152) notes in an article on the political ecology of Amazonia: "Whereas 'nature was once primarily a resource, the raw material out of which commodities may be made, the very idea of 'nature' is now a commodity, a product" (see also Escobar 1996, 1999; Lefebvre 1991). Though Rubenstein (2004, p. 153) is referring to the commodification of nature in the eco-tourism industry, similarly ironic configurations can be found in Guyana's LCDS, where it can also be said that "the struggle to 'protect' nature from economic exploitation is part of a process in which nature itself becomes a commodity ...". What is perhaps unique about the PES frontier is its attempt to redirect the commodity value of trees away from their tangible materiality and towards their intangible physiological capacities or services. In this new regime of values (Appadurai 1986), trees can be seen as producers of a commoditized service or "labor"—the rights over which can be bought and sold. In Guyana's attempt to meet development goals through PES schemes, there lies the prospect of redefining vast expanses of forest (which for many, signify underutilized space and untapped resources) into a swath of money earning, carbon storing trees—trees that not only have the capacity to develop the nation, but also to help save the world from the disastrous effects of global warming. It is taken for granted that, through efficient translation practices, all Guyanese including indigenous peoples will understand and value these arboreal capacities in the same way.

LCDS discourses draw a correlation between the preservation of standing/living trees in time-space and the long-term forward/upward development of Guyana, and likewise, between the cutting/killing of trees through deforestation and unsustainable or short-lived development.[3] I would suggest it is largely because of its irreversible spatio-temporality that this arboreal imagery has been effective in aligning development trajectories with forest conservation. Government and public media discourses that link forest conservation with the mitigation of global climate change inadvertently draw another link between deforestation in Guyana and an unsustainable mode of human existence writ large. In sum, the way in which the life trajectory of trees is understood is becoming an analogy for the sustainable socio-economic progression of the nation-state along an irreversible timeline.

The Waiwai

The peoples who collectively refer to themselves and their language as Waiwai (and often further describe themselves according to less-encompassing group names such as Hishkaryenna, Mawayenna, Xerew, Parukoto, Katuena) live in semi-permanent village communities in a predominantly forested area between southern Guyana and northern Brazil, notably along the Essequibo and Kuyuwini rivers in southern Guyana, the Jatapuzinho river in Roraima, Brazil, and the Mapuera river in Pará, Brazil (Figure 1).[4] With a total population estimated between 2,500 and 4,000, the majority of Waiwai currently reside within Brazil. Erepoimo (Big Baking Pan) and Masakenyaru (Mosquito Place), the two Waiwai communities in Guyana where my research is based, have approximately 120 and 300 residents respectively. The rhythm of everyday life in Waiwai communities is largely defined by hunting, gathering, swidden agriculture, collective work and collective meals. Numerous varieties of bitter manioc form the bulk of the Waiwai diet, supplemented by palm fruits, brazil nuts, sweet potato, banana, plantain, sugarcane, eddoe, pineapple, papaya and other fruits.[5] With the nearest town being a two to three week journey away, the Waiwai for the most part rely upon their extensive trade relations with the neighboring Wapishana, Trio and other groups throughout northeast Amazonia to obtain steel tools, shotguns, cloth and other industrially manufactured goods. The Waiwai are known regionally for their manioc grater boards, beadwork, hunting dogs, and trained parrots. Evangelical missionaries were stationed among the Waiwai from the late 1950s to 1960s, and church activities and Christian discourses are now an important part of daily life in Waiwai communities. Yet ethnographic accounts of the Waiwai since this period give the impression that these are very much circumscribed by and modified according to indigenous cosmology, lifeways, and values (Howard 2001; G. Mentore 2005; Schuler Zea 2009; L. Mentore 2010). In 2004, Conservation International-Guyana (CIG) established a protected area around Masakenyaru, and in 2005 the community gained legal title to the same area, which has since been designated a Community-Owned Conservation Area (COCA) that remains partnered with CIG. While in one sense the partnership marks a significant material and social change in the relationship between the Waiwai and the outside world, it too has been readily accommodated by the Waiwai understanding of their communities as situated at the center of an expansive and ever-transforming cosmos wrought with dangerous but desirable forms of alterity (Howard 2001). In sum, the cosmos remains for the Waiwai a dynamic multiplicity of physical and metaphysical exchanges between themselves and a vast array of other entities and energies, from animals to plants, spirits, government and NGO representatives, and the occasional researcher.

A lived environment

To help facilitate the comparative work of this paper, I invite you to attempt an interpretive shift—to conceptualize trees not as individual units evoking linear binaries and future trajectories, but as fractal entities in a recursive space-time. We might approach Waiwai ways of speaking about and engaging with trees by way of an analogy with Wagner (1991) and Strathern's (1991, 1993) notion of fractal personhood in Melanesian sociality. Neither singular nor plural but self-similar across scales, the fractal person "keeps its complexity across all scales of diminution and enlargement" (Strathern 1993, p. 49; see also Wagner 1991, p. 172). Wagner further writes that persons "[are] never a unit standing in relation to an aggregate or an aggregate standing in relation to a unit, but always an entity with relationship integrally implied" (1991, p. 163). Or as Strathern (1993, p. 49) summarizes,

> [Insofar] as persons are imagined as entities with relations integral to them, they cannot be thought of in terms of whole numbers, whether as entire units or as parts of a whole. Persons act as though they have a fractal dimensionality: however much they are divided or multiplied, persons and relations remain in proportion to each other, always keep their scale. Indeed, persons can only exist so divided or multiplied (by relations). It is as though their relationships were also themselves.

It is my understanding that for both Wagner and Strathern, the use of fractal geometry as opposed to numerical arithmetic was largely a matter of locating an analytic framework that resonated more closely with Melanesian forms of sociality, whilst also departing from the long-standing individual/society model and its implicit notion of "individuals aggregated into society" (Strathern 1993, p. 50).[6] These shifts in turn helped give way to a wave of ethnographic literature on Melanesia and Amazonia suggesting that personhood is not conceptualized as individualistic and given from the outset (e.g., at birth or through biogenetic ties), but rather, as multiplicit and emergent through social and metaphysical exchanges with other human and non-human actors (for key examples, see Strathern 1988; Gow 1991; Conklin and Morgan 1996; Viveiros de Castro 1998a, 2001; Vilaça 2002). A similar intervention into the analysis of conservation-development ideologies may be called for, given that the same forms of individuation and quantification once taken for granted in the definition of society and personhood now appear to be taken for granted in the definitions of environment through which schemes such as the LCDS are being designed and implemented.

Beyond the general political and epistemological problematics of imagining trees as individual units in a unilinear time-space, there are ethnographic grounds for suggesting the particular interpretive shift toward fractality with regard to the Waiwai environment. Firstly, in Waiwai ways of speaking there is no comfortable substitution or approximate equivalent for the English word "tree." While there is an extensive lexicon of proper names for specific varieties of trees based on their attributes and uses, there is no common noun for "tree" in the sense that we use the term to refer in a generic way to all units belonging to the so-called "tree family." Accordingly there is no plural form of such individual units—there are no "trees" in Waiwai ways of speaking and living their environment. Instead one encounters *wewe*, which in my understanding refers to an entity that might be tentatively interpreted as "wood." *Wewe* can manifest at any number of different scales: as the space beyond the village where men go to hunt; the expansive entanglement of vegetation that takes weeks to clear for planting manioc gardens; a pile of firewood; or a tiny splinter lodged inside a person's foot. These are all referred to as *wewe*, simply manifested at different scales in accordance with human motives, actions and perceptions.

Burning memories

Imagine a dozen men swinging heavy axes into towering bodies of wood in the midst of dense, tropical vegetation. Over the course of several weeks, they will clear an area sufficient for growing manioc and other produce to feed several household families for 1–2 years. Far from a plane of evenly distributed, singular entities, the space from the soft ground underfoot to the canopy looming overhead comprises an entangled meshwork of twisted branches, thorny brush and dangling, knotted lianas.[7] To the untrained eye, it can be impossible to discern what is growing up from what is growing down or along. To partition out of such a dense entanglement of life lines a space sufficient for letting in sunlight and planting human foods is no straightforward task. Injury from bodies of wood that are unpredictably pulled down by others as they come crashing down in a heaving mass is a constant possibility. Though focused on their targets, the men remain in constant communication, aware of each other's every movement.

A slow rhythm falls over the group. As they continue to chop away, their torsos comported to lunge at the same mark with every swing of their axes, their wives and daughters are busily preparing food and drink to carry to the site. When the sun approaches its zenith, they carry pots and containers along narrow pathways toward the repetitious "thud" of axes. There is the palpable sense that everyone is absorbed in the present moment of what they are doing, yet each person retains a deeper *ĩno* or knowledge/memory of the vital energies and substances their actions are eliciting. By hacking away at the life-containing wood and then pausing to enjoy the convivial pleasures of a collective meal, a metaphysical transformation of place has been set in motion.

The next vital step in this transformation is *natniyatu,* when small fires are made at the garden site and quickly spread to the dry wood strewn about. Like human bodies, animal bodies, and the bodies of material artifacts, bodies composed of *wewe* are understood to contain *katï,* fat/oil/resin or physical energy substance, and *ekatï* or spirit vitality. When the *wewe picho* or wood skin is singed away, the *katï* within is decontained and gradually absorbed into the soil, transforming it into a nourishing substance in which human foods can grow and reproduce. From time to time, the cut wood becomes so dry and heated that its *katï* becomes pressurized, making a high-pitched whistling sound and exploding out in sudden bursts. In Waiwai sensibilities, this sound signals the triumphant release of *katï.* Yet even the most effective burning does not remove all traces of *wewe* from the gardenscape.

Far from inert or meaningless, these traces continue to embody memories of the garden-making event, lending specificity to different areas of the garden and providing further energy substance for the crops. Certain varieties suitable as firewood may be deliberately spared from the flames so that their remaining *katï* can be released in the cooking of human foods at the hearth. Other remaining *wewe* may be rearranged into *wokpo* (described above). Scorched logs, stumps and even scorched patches of earth often evoke narratives about how the garden was made, who participated, and through the obligations of delayed reciprocity, who should receive portions of the food grown there. Oftentimes when I accompanied women to their gardens, we would pause during the arduous work of pulling manioc tubers as they recalled such details. We might avoid a muddy area by walking the lengths of fallen *wewe* or pass by a large stump, and my host would recall who helped to clear the area and other memorable details about the event.

As traces of *wewe* become markers, pathways, firewood, playthings for children and seats for the weary, they serve as physical and symbolic material to sustain and reproduce people and the places where they live. Yet their presence is ever-diminishing as their skins are opened up and their contents emptied out. Knowing full well that their capacity to sustain human people and places is only temporary, Waiwai eventually turn their garden sites (and even their village sites) back to the forces of an ever-encroaching entanglement of wood, vine and brush. No longer pulling the weeds around their crops, their efforts are shifted to other sites. And so continues the reciprocal exchange between the Waiwai and their lived environment.

Movement through *kanawa* space

Balance and rhythm are instantly achieved in the narrow wooden craft through deeply embodied knowledge and memories on the part of paddlers, most of whom will have been in *kanawa*, or canoe, space every day since their arrival into the human, social world. There is an impression of effortlessness despite the great amount of energy expended. The pointed tips of the blades cut quietly through the water's surface in short, swift strokes. Paddle shafts knock on gunnels as they are released from the water at a side angle. The only visible traces of the human-propelled *kanawa*—the mild wake of the boat, the rings that form as paddles cut through water—all quickly blend back into the river's seamless, glossy surface. Yet *kanawa* itself has the graphic effect of a partition. This is realized first in *netankeh*, the "opening up" of a felled tree into the form of a *kanawa*. Furthermore, the movement of *kanawa* partitions out of water a variegated array of human, social spaces.

Yet again, Waiwai ways of speaking offer no comfortable translation for our word "river," or "rivers." What we might perceive of and objectify through language as distinct rivers with formal names are seldom referred to as such by the Waiwai in their everyday speech. One encounters instead the word *tuna*. Much like *wewe, tuna* can be interpretively imagined as a fractal entity that manifests at different scales, such as *porin tuna*, the "big water" that dominates the landscape in the rainy season, and *tuunaimo*, the "huge water" that can come from all directions in the event of a severe flood. Then there are the smaller scales of *tuna* that one collects for drinking or cooking.

Waiwai refer to their riverine courses as *sama*, the same word for footpaths and tracks/prints left by people and animals. To appreciate the mutual significance of pathways and waterways as *sama* it is helpful to imagine the profound changes that occur in the landscape from the dry to rainy season. During *porin tuna* or big water, much of the low-lying land around Waiwai communities becomes submerged, concealing features of the landscape that are visible and walked upon in the dry season. This elicits changes in the bodily perspective of the human wayfarer, who can no longer walk upon the ground in certain areas and instead, glides above it in *kanawa*. Because of these seasonal oscillations, movement in *kanawa* space often entails modifying or deviating from the natural course of the main waters.

Paddlers will steer their *kanawa* straight into thickets of brush, using their bare hands, axes and cutlasses to clear fallen *wewe* or open different courses through the flooded vegetation. Initially these maneuvers struck me as impromptu decisions, but I quickly learned that we were moving along well-known waterways that are remembered season after season. What appeared to me as uninterrupted forest on either side of our *kanawa* was in fact a variegated social landscape filled with numerous *sama*. When making a long trip to another village, paddlers make frequent stops: hunting and collecting in areas known for certain animals, fruits or nuts; setting up camps in the most ideal places; and cutting select portions of *wewe* for making paddles, grater boards, and other household objects. Thus riverine wayfaring does not occur entirely on the river's surface but through a dynamic interplay between wood and water.

This interplay is perhaps most pronounced in the opening up of *wewe* into the form of dugout canoes, a process which, along with the opening up of a woman's body during childbirth, is referred to as *netankeh*. Whether we focus on bodies of wood that are opened into *kanawa*, or the flooded forest through which *kanawa* make their way, wood clearly plays a vital role in transforming water into intricate meshworks of meandering, cross-cutting lines of human sociality. Like footpaths and gardens, *kanawa* enable their paddlers to trace and modify their entangled relationships with each other and their environment season after season in accordance with the changing courses of their lives and motivations.

The *Parawa* ritual

In Waiwai understandings, all human deaths are ultimately attributable to the malevolent will—the dark shamanic violence—of another human being. Given that all human beings are understood to have the capacity to harbor such malevolence and to act upon it through shamanic violence, no human death occurs without causing serious tensions within the community. Overcome with grief, the kin of the deceased quickly set upon the question of whose malevolent intent was projected like an arrow into their body. In addition to the ritual discourses that reaffirm ties of solidarity within a community confronted with death (Fock 1963; Urban 1991), a ritual practice known as *Parawa* is also used to identify and set vengeance upon the assailant.

After the body of the deceased has been cremated in a small clearing in the forest, a male relative will collect small fragments of bone from the remains, preferably a finger or elbow joint. These fragments are inserted into a bamboo sheath along with the leaves of special plants, and the package is taken far from the village, to one of the tallest varieties of *wewe* known as *kechekere*. A small fire is made at the base of the *kechekere*, which contains *katï* that is said to "burn like gasoline" when set on fire. The fire is maintained until an inner channel has been created in the body of *kechekere* and either smoke rises from the crown or the entire body comes crashing down. With the aid of shamanic incantations and breath, the final essence of the deceased person travels through the tubular channel within the *kechekere* and into the open sky.

What I interpret as the final essence of the deceased person has been described to me as a trace of their *ekatï* or spirit vitality, and as the final remains of the knowledge and memories they once embodied. Indeed, the purpose of this "secondary cremation" is to release the deceased person's intimate memory/knowledge of their assailant and to propel it with the force of their spirit vitality into the assailant's body. When set on the invisible pathway toward vengeance, a terrible shrieking sound is said to emanate from the *kechekere*, which is heard by the assailant regardless of where s/he is located. The person undertaking the *Parawa* ritual can also do something ludicrous beside the smoldering *kechekere*, such as remove all of his clothing and run around shouting, or hold a piece of *wewe* and pretend to "sex it," causing the assailant to repeat the same behavior—typically in public and with no powers of restraint. Thus, when someone is seen behaving in such strange and disturbing ways, it is a sure sign that they are the target of *Parawa*. If successful (from the viewpoint of the relatives of the deceased), the *Parawa* ritual culminates in the assailant's swift death, and the spirit vitality once embodied by the deceased kinsperson is able to return to the anonymous pool of spirit in the swirling heavens of *kapu*. The death of the assailant makes it possible for the living kin to begin the transition from a potentially deadly state of mourning and loss to the challenging work of maintaining solidarity within the community that remains.

Conclusions

In each of these interpretive moments, whether focusing on the agricultural, navigational, or ritual life of *wewe*, it consistently operates as a mediator between the material, bodily plane of Waiwai sociality and the invisible, spiritual plane through which the former is sustained and reproduced. Waiwai cosmology, cosmography and language reflect an understanding that *there is no direct way* of accessing the spiritual truths of being. Instead, these truths emerge through an intricate meshwork of waterways and pathways, life-lines and death-lines, all mediated by fractal entities such as wood and water at different scales. What might appear from an arboreal perspective as the constant cutting and burning (i.e. irreversible killing) of trees can also be interpreted as a recognition on the part of Waiwai of inherent connections between processes of containment and decontainment, life and death.[8] The fractal and recursive existence of *wewe* is contingent upon the continual growth, release, and re-absorption of physical and spiritual vitalities between bodies of wood and others distinctively different in kind—in particular, bodies of manioc and human persons. As I have suggested, this environment may be better interpreted through a fractal geometry of scales than an arithmetic of individual units. This raises critical questions about the politics of knowing and translating an environment that is being calculated in terms of numbers of trees standing and fallen, numbers of hectares of pristine and cleared forest, numbers of gallons of carbon stored or released, numbers of dollars invested, accrued and spent.

Waiwai ways can seem roundabout and riddled with unnecessary "detours" (Schuler Zea 2009), evoking familiar judgments about the inefficiency and irrationality of indigenous peoples. As I have often witnessed, such ways can easily frustrate the sensibilities of the government surveyor or NGO representative, as they are confronted with the impossibility of navigating fractal and mediated environments as if they were (or should be) composed of linear dichotomies and direct translations. Subsequently, community consultations and other such initiatives within these environments are sometimes dismissed as "aimless" and "pointless" in and of themselves. Yet the multiple interpretive possibilities presented by Waiwai *sama* are not the result of a failure to produce or trace more efficient, more direct means to an end.

It has not been my objective in this paper to discredit the LCDS's vision of combined conservation and development goals. On many levels it can be applauded, not least of all for its attempt at a broad-based and inclusive approach and its attempt to avoid the devastating effects of large-scale commercial logging and mining. I do not doubt that some indigenous communities (many, perhaps) will choose to "opt in" to PES schemes when made available to them, and will realize some of their own socio-economic interests by engaging in such schemes. My objective has been to raise a series of questions about the stylistic and analytic approach that we are going to adopt in disciplines such as cultural and political ecology, cultural geography, and environmental anthropology as we struggle to think through this newly emerging "global ecological imaginary" and its economics and politics of knowledge, given that we are increasingly expected to fully commit to it by acting as its translators.

As others have already noted, conservation schemes tend to acknowledge indigenous cosmologies on a superficial level at best, reducing their relationships with the environment to an assemblage of "indigenous resources" or "indigenous knowledge" (Novellino 2003; Tsing 2005; West 2005, 2006; Bird-David and Naveh 2008; Velásquez Runk 2009). Using a term such as ethnographic interpretation certainly does not absolve our disciplines from the kinds of reductionism that have historically defined, and in many ways still define, Euro-American attempts to engage with "Others" through a science of translating their lifeways, environments, and values into forms that our epistemologies and economies are more prepared to deal with. Yet at the heart of engaged ethnographic practice, for me at least, is a resistance against the temptation to collapse the difficult spaces that necessarily exist between word and thing, signifier and signified. The interpretations in this paper (of the LCDS, the Waiwai, and other scholarship) are my own, and I am the first to acknowledge their imperfection. My experience of meandering (or rather, stumbling) along Waiwai ways nonetheless suggests to me the possibility of an approach to alterity that has no anxiety concerning indirect movements through imperfect interpretations along detours and roundabouts, obstructions and shortcuts, changing courses and new courses. Such excursions, whether in *kanawa* space or the space of writing, play the vital role of opening up worlds (*netankeh*, as the Waiwai might say) as opposed to collapsing them to the point where we are left with nothing but mirror images of ourselves.

Acknowledgements

Much of the research for this paper was conducted while I was a PhD student at Cambridge University, where I was the William Wyse Student of Social Anthropology from 2004–2007. In addition to the Wyse Studentship, my field-work was funded by a grant from the Anthony Wilkins Fund. Earlier versions of this paper were presented at the University of Mary Washington in January 2010, and at the "Amazonian Geographies" session at the Annual meetings of the Association of American Geographers in April 2010. I thank both audiences for their instructive questions and comments. I am grateful as always to my husband, George, for his support and encouragement throughout the many drafts of this paper. Above all, I am indebted to the people of southern Guyana, who continue to graciously open their homes and lives to us and our daughter, Kamina.

Notes

1. Unfortunately there is not space here to properly review the well-established discussions and debates in anthropology concerning translation and the notion of "culture as a text". For key references, see Gellner (1970), Geertz (1973), Asad (1986). The 2002 issues of *Anthropology News* also contain an informative series of discussions under the heading "Problems of Translation."
2. Schuler-Zea (2009) offers a similar account of Waiwai ways of knowing the environment as indirect and mediated. Though translation is one of her main theoretical concepts, she refers not to translation-proper but to "improper translation," a concept taken from theorists who problematize conventional assumptions about translation, such as Benjamin (2004), Asad (1986) and Derrida (1976). In her "improper" approach to translation, Schuler Zea's argument would seem to support the concept of interpretation used in this paper.
3. On symbolic associations between the longevity of trees and human social histories, see Rival (1998).
4. The most comprehensive anthropological studies of the Waiwai to date are G. Mentore (2005) and Howard (2001). Earlier accounts by Fock (1963) and Yde (1965), though driven by a "salvage anthropology" approach, are valuable sources on Waiwai cosmology and material culture. For syntheses of the ethnography of the Guianas region, see Rivière (1984) and Overing (1981).
5. The term "varieties" is used instead of "species" to avoid suggesting that the Waiwai classify plants and trees based on the same criteria as the Western taxonomic classification of species.
6. For an earlier use of fractal principles in anthropology see Haraway (1991).
7. Lefebvre (1991) uses the term meshwork to refer to reticular patterns or webs of lines made by the movements of animals and people in the environment. Here

it is meant to provide an alternative spatial imagery to the "network" composed of fixed points or centers. On this distinction, see Ingold (2007, p. 80).

8. This is an important point given the tendency in conservation discourses to focus disproportionately on indigenous swidden or "slash and burn" agricultural techniques as forms of deforestation and carbon emissions in comparison to much larger scale and less sustainable forms of commercial agriculture and ranching.

References

Appadurai, A., 1986. *The social life of things: commodities in cultural perspective.* Cambridge: Cambridge University Press.

Asad, T., 1986. The concept of cultural translation in British social anthropology. *In*: J. Clifford and G. Marcus, eds. *Writing cultures: the poetics and politics of ethnography.* Berkeley, CA: University of California Press, 141–164.

Balée, W., 1994. *Footprints of the forest: Ka'apor ethnobotany – the historical ecology of plant utilization by an Amazonian people.* New York: Columbia University Press.

Benjamin, W., 2004. The task of the translator. *In*: L. Venuti, ed. *The translation studies reader.* London: Routledge, 15–25.

Bird-David, N. and Naveh, D., 2008. Relational epistemology, immediacy, and conservation: or, what do the Nayaka try to conserve? *Journal for the Study of Religion, Nature and Culture,* 2 (1), 55–73.

Conklin, B. and Graham, L., 1995. The shifting middle-ground: Amazonian Indians and eco-politics. *American Anthropologist,* 97, 695–710.

Conklin, B. and Morgan, L., 1996. Babies, bodies and the production of personhood in North American and a Native Amazonian society. *Ethos,* 24 (4), 657–694.

Deleuze, G. and Guattari, F., 2007. *A thousand plateus: capitalism and schizophrenia.* Brian Massumi, trans. Minneapolis: University of Minnesota Press.

Derrida, J., 1976. *Of grammatology.* Spivak, G., trans. Baltimore, MD: Johns Hopkins University Press.

Descola, P., 1996. *In the society of nature: a native ecology in Amazonia.* Cambridge: Cambridge University Press.

Ellis, D. and West, P., 2004. Local history as indigenous knowledge: aeroplanes, conservation and development in Haia and Maimafu, Papua New Guinea. *In*: A. Bicker, P. Stillitoe, and J. Pottier, eds. *Investigating local knowledge: new directions, new approaches.* London: Ashgate, 105–127.

Errington, F. and Gewertz, D., 2001. On the generification of culture: from blowfish to Melanesian. *Journal of Royal Anthropological Institute* (new series), 7 (3), 509–525.

Escobar, A., 1996. Constructing nature: elements for a post-structuralist political ecology. *In*: R. Peets and M. Watts, eds. *Liberation ecologies: environment, development and social movements.* New York: Routledge, 44–68.

Escobar, A., 1999. After nature: steps to an anti-essentialist political ecology. *Current Anthropology,* 40 (1), 1–30.

Fabian, J., 1983. *Time and the other: how anthropology makes its object*. New York: Columbia University Press.

Fock, N., 1963. *Waiwai: religion and society of an Amazonian tribe*. Copenhagen: National Museum of Copenhagen.

Geertz, C., 1973. *The interpretation of cultures: selected essays*. New York, NY: Basic Books.

Gellner, E., 1970. Concepts and societies. *In*: B. Wilson, ed. *Rationality (key concepts in the social sciences)*. Oxford: Blackwell, 18–49.

Government Information Agency of Guyana, 2010. *Joint concept note on REDD+ cooperation between Guyana and Norway* [online]. Georgetown, Guyana. Available from: http://www.gina.gov.gy/Joint%20Concept%20Note.pdf [Accessed June 2010].

Gow, P., 1991. *Of mixed blood: kinship and history in Peruvian Amazonia*. Oxford: Oxford University Press.

Haraway, D., 1991. *Simiens, cyborgs and women: the reinvention of nature*. New York: Routledge.

Harper, J., 2002. *Endangered species: health, illness and death among Madagascar's people of the forest*. Durham, NC: Carolina Academic Press.

Harrison, R., 1992. *Forests: the shadow of civilization*. Chicago, IL: University of Chicago Press.

Hennessy, A., 1978. *The frontier in Latin American History*. Bristol, UK: Edward Arnold.

Hennessy, L., 2003. Silencing cultural land: Amerindians, the environment, and cooperative socialism in post-colonial Guyana, 1956–1986. Paper presented at the International Conference on the Forest and Environmental History of the British Empire and Commonwealth, University of Sussex, March 19–21.

Howard, C., 2001. *Wrought identities: the Waiwai expeditions in search of the "unseen tribes" of northern Amazonia*. Thesis (PhD). University of Chicago.

Ingold, T., 2007. *Lines: a brief history*. London: Routledge.

Jagdeo, President Bharrat, BBC News website 2008. *Viewpoint: why the West should put money in the trees* [online]. Available from: http://news.bbc.co.uk/2/hi/science/nature/7603695.stm [Accessed June 1, 2010].

Kaieteur News Press, 2010, Kaieteur News website. *Sukhai questions translating LCDS into Amerindian languages* [online]. Available from http://www.kaieteurnewsonline.com [Accessed June 2010].

Lefebvre, H., 1991. *The production of space*. Oxford: Basil Blackwell.

Little, P., 2001. *Amazonia: territorial struggles on perennial frontiers*. Baltimore: Johns Hopkins University Press.

Maranhão, T., 2003. The politics of translation and the anthropological nation of the ethnography of South America. *In*: T. Maranhão and B. Streck, eds. *Translation and ethnography: the anthropological challenge of intercultural understanding*. Tucson, AZ: University of Arizona Press, 64–84.

Maynard, K. and Paquin, M., 2004. Payments for environmental services: a survey and assessment of current schemes. Unisféra International Centre, for The Commission of Environmental Cooperation in North America. Montreal, September 2004. [online] Available from: http://www.cec.org/Storage/56/4894_PES-Unisfera_en.pdf [Accessed March 2010].

Mentore, G., 2005. *Of passionate curves and desirable cadences: themes on Waiwai social being*. Lincoln: University of Nebraska Press.

Mentore, L., 2010. *Trust and alterity: Waiwai analyses of social and environmental relations in Southern Guyana*. Thesis (PhD). University of Cambridge.

Murray Li, T., 2007. *The will to improve: governmentality, development and the practice of politics*. Durham: Duke University Press.

Novellino, D., 2003. Contrasting landscapes, conflicting ontologies: assessing environmental conservation on Palawan Island (the Phillippines). *In*: D.G. Anderson and E. Berglund, eds. *Ethnographies of conservation: environmentalism and the distribution of privilege*. New York: Berghahn, 171–188.

Office of the President, Government of Guyana website 2010. *Guyana's low carbon development strategy* [online]. Available from: http://www.lcds.gov.gy [Accessed 1 June 2010].

Overing, J., 1981. Review article: Amazonian anthropology. *Journal of Latin American Studies*, 13 (1), 151–164.

Posey, D., 2007. Indigenous management of tropical forest ecosystems: the case of the Kayapo Indians of the Brazilian Amazon. *In*: M. Dove and C. Carpenter, eds. *Environmental anthropology: a historical reader*. Malden, MA: Wiley-Blackwell, 89–101.

Rival, L., 1998. *The social life of trees: anthropological perspectives on tree symbolism*. Oxford: Berg.

Rivière, P, 1984. *Individual and society in Guiana: a comparative study of Amerindian social organization*. Cambridge: Cambridge University Press.

Rubenstein, S., 2004. Steps to a political ecology of Amazonia. *Tipiti: Journal of the Society for the Anthropology of Lowland South America*, 2 (2), 131–176.

Schuler Zea, E., 2009. Metaphoric detours and improper translations in the double field of Waiwai anthropology. *In*: N. Whitehead and S. Aleman, eds. *Anthropologies of Guyana: cultural spaces in northern Amazonia*. Tucson, AZ: University of Arizona Press, 207–221.

Smith, N., 1990. *Uneven development: nature, capital and the production of space*. Athens, GA: University of Georgia Press.

Strathern, M., 1988. *The gender of the gift: problems with women and problems with society in Melanesia*. Los Angeles, CA: University of California Press.

Strathern, M., 1991. *Partial connections*. Savage, MD: Rowman and Littlefield.

Strathern, M., 1993. One-legged gender. *Visual Anthropology Review*, 9 (1), 42–51.

Sukhai, Minister Pauline, Indigenous Peoples Issues and Resources website, 2010. *Indigenous peoples rights always accorded top priority according to Guyana Minister* [online]. Available from: http://www.indigenouspeoplesissues.com [Accessed May 2010].

Tsing, A., 2005. *Friction: an ethnography of global connection*. Princeton: Princeton University Press.

Tuan, Y-F., 1977. *Space and place: The perspective of experience*. Minnesota: University of Minnesota Press.

Urban, G., 1991. *A discourse-centered approach to culture: native South American myths and rituals*. Austin: University of Texas Press.

Velásquez Runk, J., 2009. Social and riverine networks for the trees: Wounaan's riverine rhizòmic cosmos and arboreal conservation. *American Anthropologist*, 111 (4), 456–467.

Vilaça, A., 2002. Chronically unstable bodies: reflections on Amazonian corporealities. *Journal of the Royal Anthropological Institute*, 11 (3), 445–464.

Viveiros de Castro, E., 1998a. Cosmological deixis and Amerindian perspectivism. *Journal of the Royal Anthropological Institute*, 4 (3), 469–88.

Viveiros de Castro, E., 2001. GUT feelings about Amazonia: potential affinity and the construction of sociality. *In*: L. Rival and N. Whitehead, eds. *Beyond the visible and the material: the Amerindianization of society in the work of Peter Rivière*. Oxford: Oxford University Press, 19–44.

Wagner, R., 1991. The fractal person. *In*: M. Godelier and M. Strathern, eds. *Big men and great men: the personification of power*. Cambridge: Cambridge University Press, 159–173.

West, P., 2005. Translation, value, space: theorizing an ethnographic and engaged environmental anthropology. *American Anthropologist*, 107 (4), 632–642.

West, P., 2006. *Conservation is our government now: the politics of ecology in Papua New Guinea*. Durham, NC: Duke University Press.

Yde, Y., 1965. *Material culture of the Waiwai*. Copenhagen: National Museum of Copenhagen.

Zerner, C., 2003. Moving translations: poetics, performance and property in Indonesia and Malaysia. *In*: C. Zerner, ed. *Culture and the question of rights: forests, coasts and seas in Southeast Asia*. Durham, NC: Duke University Press, 1–23.

Redefining identities, redefining landscapes: indigenous identity and land rights struggles in the Brazilian Amazon

Omaira Bolaños, Ph.D.

Rights and Resources Group, Washington, D.C., USA

Amazonian ecological and cultural landscapes have been cause for debate between environmental and development agencies that have attempted to re-configure them into conservation regions and frontier areas for agricultural expansion. However, it has also been the diverse Amazonian peoples, through their socio-political dynamics, rights struggles, and reactions to state control of space and resources, that have re-defined their identities and landscapes. This article examines the political struggle for the revitalization of indigenous ethnic identities and the recognition of territorial rights of the Indigenous Council of the Lower Tapajós-Arapiuns (CITA) in the state of Pará, Brazil. CITA's claims became visible in 1998 within disputes between globalizing development and conservation strategies that questioned the legitimacy of indigenous land rights claims and identity. Findings indicate that essentialized notions of indigenous peoples and the Amazonian landscape have been appropriated by non-indigenous actors to strategically argue against CITA's rights claims. I argue that the changes perceived in both the Amazonian rainforest and indigenous peoples' socio-cultural formations have become a discursive strategy to legitimize the expansion of agricultural production into the secondary forest, and a way to deny the rights of indigenous peoples.

Introduction

In the article "The Temporality of the Landscape," Ingold and Bradley (1993) assert that life is a process that involves the formation of the landscape in which people live. The landscape, they state, "is constituted as an endured record of—and testimony to—the lives and works of past generations who have dwelt within it" (Ingold and Bradley 1993, p. 152). Landscape becomes part of what people are through the everyday experiences of those who dwell in it, work it, struggle on it, and move along it. The way people create meaningful relationships with places, and the significance of this relationship for the development of the conception of the self, is a key issue in the association between identity and land. Yet, people and their landscape change over time through internal and external forces that contribute to the re-definition of space and identities. Scholars argue that the mental, spiritual, and intellectual development of the individual self is the product of the interconnectivity of the self and society with the Earth (Appiah-Opoku 2007).

Indigenous peoples have been historically linked to the land as their main source of survival, and as an essential element for the preservation of their culture and distinctive ethnicities. In international and national legislation, land and resources constitute a fundamental base for the protection of indigenous peoples' rights (United Nations 2007). In indigenous political activism, the close connection with ancestral lands and forest resources constitutes a key component of the discursive appeal to demonstrate an enduring record of their territorial rights.

Yet, forests are important socio-economic and cultural components of indigenous and other forest-dependant peoples' lives. Indigenous peoples have for centuries re-shaped and managed the forest ecosystems on which they depend to sustain life and much of what is defined as primary forest is in fact a socio-historical and cultural landscape (Posey and Baleé 1990; Denevan 1992; Gómez-Pompa and Kaus 1992; Baleé 2003; Heckenberger 2007). The scientific community has established different categories of forest types that define their economic and biodiversity value, such as primary (or pristine), and secondary forests. These categories are defined in terms of the level of impact of human activity, in which primary forest is presumed to lack, or has been little affected, by human activity, whereas secondary forest has been highly altered (ITTO 2002).[1] Rocheleau et al. (2001) argue that the primary/secondary forest dichotomy that still dominates the scientific literature disregards the biological, cultural, and economic values of forest gardens and secondary forest systems that support the livelihoods of many forest-dependant people around the world. More critically, this dichotomy perpetuates the idea of forest-dwelling communities as antagonistic, and endangers indigenous peoples' rights to access and control their territories and forest resources. At the same time, it perpetuates the idea of forest as fixed and unchanging (Gómez-Pompa and Kaus 1992).

In this article, I am concerned with the revitalization of indigenous identity through processes of territorial rights struggles in the lower Tapajós-Arapiuns region in the state of Pará, Brazil. Since the late 1990s, peoples from the region have been claiming recognition of their indigenous status and the demarcation of their territories. My intent is to show how identity and territorial rights claims have spawned discursive strategies among non-indigenous actors to deny the rights of the peoples of the lower Tapajos-Arapiuns region. I show how different contesting discourses against indigenous peoples' claims stem from the competing interests that vie for control of Amazonian forest landscapes. Specifically, I ask: How do peoples of the lower Tapajos-Arapins articulate their indigenous identity to land rights claims? What are the discursive strategies that non-indigenous groups use to contest indigenous peoples' claims? How do non-indigenous people interpret the changes perceived in the indigenous peoples' social formation and their landscape?

By claiming territorial rights, the people from the lower Tapajos-Arapiuns region have revitalized the memories and history that reinforce their sense of indigenousness. Organized through the Indigenous Council Tapajos-Arapiuns (CITA), 38 communities[2] mobilized to claim rights and contest state definitions of their own identity and landscape that affected control over their territories. However, this process of identity revitalization and land claims became highly contested and wrapped up in ongoing environmental and economic disputes over Amazonian landscapes and resources. These disputes include conservation efforts to protect and promote sustainable management of the Amazonian tropical rainforest, and economic development related to the paving of the highway BR-163 (which links the Mato Grosso soybean industry to the Santarém, Pará port). Advocates of economic development point out that the highway is needed to consolidate Brazil's soy production in the international market and bring economic prosperity to the region. Environmentalists, on the other hand, assert that the highway construction and soybean agriculture are increasing deforestation, global warming and land grabbing. Within this context of competition for the Amazonian landscape, indigenous peoples' land struggles were not only blurred, but also highly scrutinized and questioned because of the apparent acculturation of CITA members—mainly their loss of language, religion and traditional indigenous clothing.

The people from the lower Tapajos-Arapiuns region are commonly defined as *ribeirinho*, people who live along rivers, or as *caboclo*[3]—a mixture of whites (Portuguese), Indians, and African slaves that originated during the colonial period (Moreira 1988). In traditional literature regarding the indigenous peoples of the Amazon, the *caboclo* is portrayed as the antagonist of indigenous culture and as the agent of radical changes within Amazonian landscapes and society (Wagley 1976; Galvão 1979; Ross 1978; Parker 1989). In other words, the *caboclo* was conceived as a stage in a continuum of socio-cultural change that marks the disintegration of indigenous cultures (Bolaños 2010). However, in what is considered a process of reversal, numerous groups defined as *caboclo* are now recuperating their indigenous culture and identity. In different parts of the country, many indigenous peoples of mixed descent[4] have become political activists who struggle for the recognition of their rights (Oliveira 1999; Warren 2001; Ioris 2005). Guzman (2006) shows how, in the Negro

River region, indigenous populations experienced a complex process of cultural interchange and miscegenation before the arrival of Europeans, and continue to do so by incorporating other social groups. However, people from this region are perceived as acculturated, the product of changes brought by Europeans. Scholars argue that the marginalization of the *caboclo* as a subject of Amazonian research has made the historical, cultural and economic development of the *caboclo* culture invisible (Nugent 1993; Harris 1998). Recent studies on the invisibility of the Amazonian peasantry highlight the complex historical processes that gave origin to *caboclo* society, and the political, social and environmental contexts in which the *caboclo* identity and landscape have been re-shaped (Zarin *et al.* 2001; Raffles 2002; Murrieta and Winkler-prins 2003; Nugent and Harris 2004; Adams *et al.* 2009). For instance, Raffles (2002) explains that *ribeirinho* populations have historically transformed the Amazonian fluvial landscape by the very economic and intimate interactions between people and their landscape. Based on the case of the *igarape* Guariba in the Brazilian Amazon, Raffles (2002) shows how a simple stream was transformed into a wide and intricately ramified river. These changes not only helped expand economic activity but also contributed to the formation of group identity and sense of belonging.

In this paper, I intend to contribute to the exploration of emerging identities and landscapes in Amazonian cultural geography through the analysis of the complex process of territorial and identity re-definition of peoples not previously recognized as indigenous by the state. My analysis focuses on ethno-political action[5] and explores it in two ways: as the use of historical and socio-political discourses that support the rights claims of indigenous peoples, and as the use of competing arguments against these claims by other social agents (e.g., the general public, and governmental and economic development actors). I argue that non-indigenous actors use to their advantage their own conceptions of indigenousness to question the legitimacy of indigenous peoples' rights claims. In this context of developmental and environmental disputes, non-indigenous actors have appropriated essentialized notions of Amazonian indigenous peoples and their forest landscape as a means to question peoples' identity and land claims. I highlight the parallel between the interpretation given to the changes perceived in the Amazonian rainforest ecosystems, and in indigenous peoples' social formation. I explore how the notion of change is appropriated to legitimize the expansion of a more profitable activity (i.e., soy cultivation and infrastructural improvement) into Amazon forest areas considered to have changed too much to contain biodiversity value. Similarly, I show how the notion of change is used to question the legitimacy of indigenous peoples of mixed descent, claiming they have changed too much to be "authentic" Indians.

Methods

I conducted research between 2004 and 2007 in two areas: the indigenous territories along the lower Arapiuns River and other communities along the Tapajós River; and in the city of Santarém, base of the CITA office, where members of all ethnic groups periodically gathered. To undertake the field research, I used ethnographic techniques, including 51 semi-structured interviews, 13 life histories, and participant observation. I also collected information from secondary sources such as the Santarém city newspaper, and institutional letters and reports, to support analysis and understand the socio-economic context.

For the purposes of this paper, I draw analyses from life histories of indigenous leaders and recorded observations from the May-July 2006 public demonstrations that took place in the city of Santarém to protest deforestation in the Amazon. These public demonstrations revealed confrontational discourses regarding the Amazonian forest landscape and indigenous claims. On one side there were civil society groups, including CITA, national and international environmental NGOs to protest expansion of agribusiness frontier, the establishment of a soybean export terminal in Santarém, and deforestation caused by soybean production. On the other side, local and regional soybean farmers and merchants congregated to defend the economic prosperity brought by the soybean industry. I also used materials collected from local newspapers that reported on the May-July 2006 public demonstrations.

In this study, I used purposive sampling, which is useful in the selection of informants in a deliberative, predetermined, and non-random sample (Bernard 2002). The method was used for the purpose of finding individuals who self-identified as indigenous, belonged to the communities claiming land demarcation, had CITA membership, and were permanent residents of the communities. The life histories were developed with key members, such as the *caciques*, people designated as important within the communities, and CITA leaders. All interviews were tape recorded with prior participant permission, and verbatim transcriptions were produced to conduct data analysis. I participated in indigenous peoples' daily activities in their communities, in meetings, and special events carried out by CITA. I recorded my observations as field notes and described activities, people's interactions, and settings through narrative (Merriam 2002; Denzin and Lincoln 2003). Participant observation helped me situate my research in social context and helped reveal the intricate relationships among the different actors.

I use grounded theory methodology to analyze data from interviews and field notes. This method is useful for understanding people's experience in a detailed manner focusing on meaning (Strauss and Corbin 1998; Charmaz 2006). Given that I was dealing with a very critical issue in terms of the emotions, passions, and political controversy that it generates, grounded theory helped me to keep the richness of concepts and sentiments expressed in the individual stories. A grounded theory approach allows for the identification of themes and concepts that emerge from peoples' narratives. It also preserves individual voices and inter-viewers' experiences and perspectives, on such issues as identity and landscape re-definition. I followed the guidelines and principles developed by Charmaz (2006) on grounded theory, which allowed me to develop a descriptive analysis of what informants expressed. It helped me to group and contrast data according to conceptual content, and to find relation-ships among different emerging themes.

Background

The indigenous movement of the lower Tapajós-Arapiuns region in the state of Pará emerged within a context of long-term negotiations between the government and local communities to resolve the conflict of the overlapping of local peoples' lands with those of the Flona Tapajós Forest Reserve (Ioris 2005). The indigenous peoples from the region started mobilizing in 1997, and in 1998 the community of Takuara, located within the Flona Tapajós Forest Reserve, claimed legal recognition as indigenous Munduruku, and the demarcation of their land, by FUNAI (Brazilian National Indian Foundation) (Vaz 2004; Ioris 2005). The movement rapidly spread through the lower Tapajós-Arapiuns rivers region and 12 different indigenous groups, represented by CITA, started claiming their land rights: Arapium, Tapajó, Jaraki, Munduruku, Arara-Vermelha, Apiaka, Tapuia, Tupinambá, Borari, Maitapú, Cara Preta, and Camaruara (Bolaños 2010).

In 2008, FUNAI organized Technical Groups (TG) to carry out anthropological and environmental studies to elaborate the proposals for land demarcation in five indigenous lands of the region: Borari/Alter do Chão, Maró, Aningalzinho, Cobra Grande, and Escrivão (PPTAL 2008). These demarcation processes are still in the second stage of the administrative proceeding, which includes analysis of the contesting claims that non-indigenous peoples have on the land demarcation proposals (Lima 2009). In a communication dated 29 October 2009, FUNAI announced the approval of the anthropological report on the demarcation of the Munduruku-Taquara indigenous land that had been developed since 2003 (FUNAI 2009).

As will be explored in the next sections, within the complex emerging landscape of western Pará, CITA's claims became a contentious issue in the negotiations amongst the diverse actors. Responses to their claims ranged from support to arguments that denied indigenous peoples' land rights. One of the main arguments against CITA has been the apparently dramatic process of cultural change and racial mixture experienced by the people it represents. Opponents of the indigenous movement's claims argue that CITA members can no longer be considered indigenous peoples because they have lost their language and their most traditional customs and religion (Amazônia 2009).

Emerging Amazonian landscapes

Although Amazonia constitutes the world's largest tropical rainforest in which long-term international and national conservation efforts are concentrated, deforestation produced by economic development projects continues to be one of the major threats to its biodiversity value (Fearnside 2007). The state of Pará, with 1,247.703 km^2, represents 15% of Brazilian territory. Of the total area of the state, 73% is covered by forests (Veríssimo et al. 2002). Pará is one of the top five timber-producing states of the Brazilian Amazon, containing 24 logging area centers.[6] In the central and western part of the state, timber production has intensified since the 1990s due to the expected paving of the BR-163 highway, which connects the cities of Cuiabá and Santarém (Veríssimo et al. 2002). The BR-163 was constructed in 1973 as a dirt road, and its poor condition impeded the massive influx of migrants and investment. The expectation that it will be paved has already prompted migration, the influx of soybean producers into the region, and increased deforestation (Brown et al. 2005; Soares-Filho et al. 2006; Fearnside 2007; Vera-Diaz et al. 2009). Soybean production has become the driving force behind the development of the agricultural industry. Vera-Diaz et al. (2009) show that annual soybean production between 2000 and 2005 grew from 9 to 20 million tons, and that soybean acreage increased from 31,000 to 70,000 km^2 in the Amazon basin. The profitability of the soybean industry has motivated many Brazilian states to boost their economies through the conversion of subtropical and tropical areas to soy production, and to support the improvement of the transportation network (Soares-Filho et al. 2006; Vera-Diaz et al. 2009).

Conflicts over land and resources have intensified and become prominent forces for deforestation, violence and impunity (Simmons et al. 2007). Some studies suggest that the violent land struggles in Pará are the result of a historical process of land ownership concentration (Sauer 2005). Corruption, and lack of implementation of governmental regulations related to environmental laws and human rights, has contributed greatly to this violent land conflict (Sauer 2005). On the other hand, the clearing of new forest areas has become the most effective way to establish control and claim land tenure rights (Fearnside 2007).

The establishment of forest reserves and extractive reserves (RESEX) became a crucial environmental strategy to contain the deforestation caused by road development and agricultural expansion, while at the same time promoting sustainable land and forest resource use (Figure 1). The RESEX are a part of the National System of Conservation Units

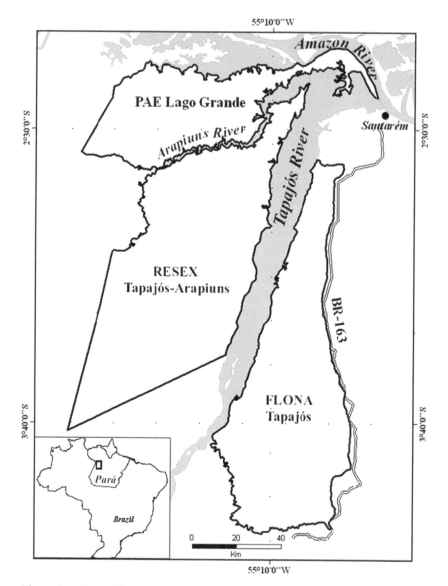

Figure 1. General location of the study area, including the Conservations Units: Flona Tapajós Forest Reserve, Tapajós-Arapiuns RESEX, and Lago Grande PEA, and the BR-163.

(SNUC)—Law no. 9.985, 2000—which includes forest reserves (FLONA), areas of environmental protection, areas of ecological relevance (ARIE), fauna reserves (FR), and sustainable development reserves (RDS). The concept of RESEX first emerged as an innovative bottom-up policy from rubber-tappers' struggle to ensure land rights over areas of forest occupied by generations. A RESEX establishes a territory and regulates access to natural resources by populations that traditionally and culturally depend on extractivism for subsistence. The first extractive reserve was created in 1990 in Acre in the Alto Jurua (Barreto 2009).

International and national environmental NGOs, the government, and the economic sector have become involved in the intricate issue of finding common ground between economic development and conservation initiatives in the lower Tapajós-Arapiuns region (Greenpeace 2008). A working group of environmental organizations and representatives, from Mato Grosso and Pará's 84 municipalities, was created to debate the direct and indirect impacts of the highway's paving, and alternative ways to construct a regional sustainable development model (IPAM 2004). Among other things, this group prioritized the creation and consolidation of protected areas, the regularization of existing agrarian reform settlements, support for family agriculture and sustainable activities, and the improvement of infrastructure services (IPAM 2004). Between 2004 and 2006, 23 million hectares of protected areas were created, including those along the pathways to the agricultural frontier (Nepstad *et al.* 2008).

When the Flona Tapajós Reserve was created in 1974, attempts were made to displace the local population, and restrictions on forest resource use were imposed on those inhabiting the reserve area (Ioris 2005). The creation of the forest reserve established a new reality for the area's inhabitants. State intervention regulated resource use and land occupation, and redefined their territorial rights and identities to that of "traditional people." The concept of traditional people appeared at the end the 1980s in the national debates around human presence in protected areas. As many social groups suffered the effects of restrictions imposed over access to and use of resources, as well as forced displacement, supporters of the rights of protected areas' residents mobilized to demonstrate that these residents did not represent a threat to biodiversity conservation (Barreto 2009). Several attempts at defining the rights of traditional people within protected areas culminated with the creation of Decree 6040 of 7 February 2007, which established the National Policy for Sustainable Development of People and Traditional Communities (Creado *et al.* 2008). This Decree made the existence of traditional communities within the protected areas legal, while the traditional populations, in return, commit to make sustainable use of natural resources. The Decree also provides a definition of traditional people as those groups who recognize themselves as culturally distinct, maintain their own forms of social organizations, and occupy and use territories and natural resources for their cultural, social, religious, and economic reproduction.

Previous to Decree 6040, attempts in defining who should be considered as traditional people were vetoed, because definitions were considered too broad (Creado *et al.* 2008). However, within the current definition of traditional people are included a diversity of social groups, among them descendants of Afro-Brazilian communities, peasants, *ribeirinhos*, and *caboclos*. As Barreto (2009) argues, in practice, the concept represents an oversimplification of the socio-cultural reality that does not distinguish diverse forms of social, cultural, and territorial organization. It constitutes a classificatory homogenizing definition constructed from the outside and imposed on local peoples. Barreto also argues that the definition of traditional people constitutes an idealization that depicts these social groups as frozen in time; that is, as if the conditions of their existence and way of life were not subject to change (Barreto 2009).

The reaction to the ascription to this new identity in the lower Tapajós was the emergence of an indigenous movement to reinstate their ethnic identities, claim the demarcation of their lands, and the right to exercise a particular way of life as indigenous peoples (Ioris 2005). The regional mobilization to defend their land rights opened the opportunity to recover the history and collective memories that made this landscape a material and symbolic record of their indigenous descent. Vaz (2008) asserts that discussions about land rights as traditional people in the lower Tapajós-Arapiuns facilitated collective reflection on the origin of this definition, and contributed to a process of reevaluation and reconstruction of their indigenous ethnicities. This increased their motivation to claim indigenous identity and to invest a new and positive value in the meaning of being indigenous (Bolaños 2010).

In 1998, the Tapajós-Arapiuns Extractive Reserve (RESEX) was created covering an area of 647,000 hectares. This increased the conflict due to even more overlap between the lands being claimed by indigenous peoples and the legally recognized rights of "traditional" people residing within the new RESEX area (see Figure 2). The governmental mobilization for the creation of the Tapajos-Arapiuns RESEX coincided with the formation of CITA and their claims for land demarcation in the same area where the RESEX Tapajos-Arapiuns was already established. Indigenous peoples living in the RESEX Tapajós-Arapiuns and other indigenous communities located on the left bank of the Arapiuns River mobilized to claim their land rights and assert their indigenous ethnic identities.

Several regional meetings took place among the various social groups inhabiting the RESEX, government officials, IBAMA (Brazilian Institute of Environment and Renewable Natural Resources), and NGOs to discuss different types of land rights in the area, and resolve conflicts between people who self-identified as indigenous, and those who did not (CITA 2006). Currently, there are 20 indigenous communities of the extractive reserve claiming indigenous land demarcation.[7] However, since not all the communities within the RESEX claim their indigenous land rights and identity, different types of conflicts were generated. Some conflicts

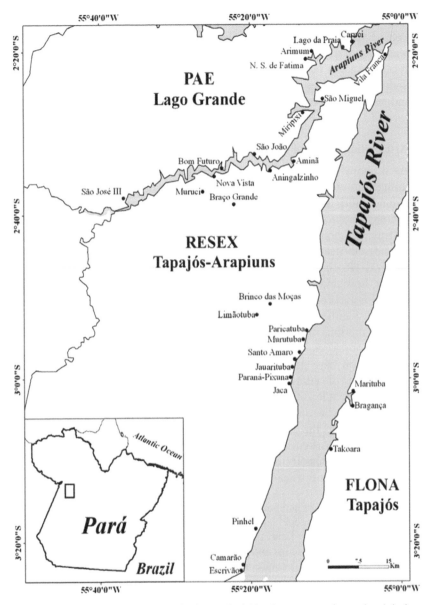

Figure 2. Indigenous communities located within the conservation units claiming recognition and demarcation of indigenous lands.

were related to the acceptance by non-indigenous communities of the coexistence of indigenous peoples with demarcated land within the RESEX area. Others were related to representation in the RESEX administration unit board. Agreements of mutual respect and recognition of the legitimacy of indigenous and non-indigenous organizations and land rights were signed (Pontes 2003). In 2005, the Lago Grande Agro-Extractive Settlement Project (PAE), with an area of 2,503.44 m^2, was created (Saude e Alegria 2007). The area of the Lago Grande PAE is located on the left bank of the lower Arapiuns River and also includes eight communities that had previously requested demarcation of their indigenous territories.

Articulating land and identity

As in many other cases around the world, claims for the recognition of indigenous status in the lower Tapajós-Arapiuns are tied to struggles to defend and secure ancestral territories. The concept of territory encompasses symbolic and material meanings expressed through culture, religion, spiritual sites, memories, forest resources, water, etc., considered necessary for indigenous peoples' cultural and economic survival (Rocheleau *et al.* 2001; Stavenhagen 2005). Perreault (2001) shows how indigenous federations of the Ecuadorian Amazon have linked issues of resource access with the concept of territory and cultural identity to their discourse of political autonomy. Dramatic transformations in the last three decades in the Ecuadorian Amazon due to processes of oil production, agrarian reform, and large-scale colonization have generated indigenous peoples' mobilizations to secure recognition of ethnic nationalities based on territorial and political rights (Perreault 2001).

CITA's territorial claims are based on the concept of original home-lands. That is, the land that has historically provided the sources of livelihood, subsistence, and the basis of their existence as distinctive people, as expressed in the quote below from an interview with an Arapium indigenous leader[8] from the community of Caruci:

> The main right of the *índio* (indigenous person) is the land, which is our original right to our lands. We hold the right to work and to live permanently in our land, the land of our ancestors. When I die, this land will be for my grandsons and granddaughters, and so on. Then, we keep the land for our people, because here, we all are *índios*. (Interview by author, 20 July 2004)

Nobre (2002) asserts that, since indigenous rights to land are founded on the notion of indigenous people as the first inhabitants of the Brazilian territory, land rights are practically a pre-existing right, that is, prior to law. The demarcation of indigenous land by the state means that the state recognizes pre-existing rights over a particular territory, and these rights establish the extent of the territory as well as rights of ownership and use. In contrast, Oliveira (1998) explains that the application of the concept of original rights does not occur because of the legal recognition by the state (nor is it revoked by non-recognition). He asserts, "The concept is formed from the continued existence of indigenous groups that have a connection with pre-Colombian populations, consider themselves as ethnically distinct groups, and maintain certain ancestral traditions" (Oliveira 1998, p. 45).

The 1988 Brazilian Constitution explicitly recognized the rights of indigenous peoples to maintain their cultural traditions. The constitution also recognizes that the lands traditionally occupied by indigenous peoples are imbued with cultural meaning and are "lands considered vital for the preservation of environmental resources necessary for their welfare and physical and cultural reproduction according to their traditional uses and customs" (Mendes 2005, p. 15). Scholars allege that although Brazilian law on indigenous lands is consistent with the legal system, it hides a more basic right of indigenous peoples—the right to territory. What Brazilian law guarantees is indigenous peoples' right to permanent possession and exclusive usufruct of natural resources, but the state has not renounced control of indigenous lands and territories (Ramos 2003).

According to FUNAI, the number of indigenous lands recognized and demarcated has tended to increase as more new indigenous groups are known. Some of these indigenous groups constitute those defined by FUNAI as "resurgent ethnicities," that is, those indigenous groups that "although presumed extinct, reappear, asserting and claiming recognition of their indigenous identity and territorial rights" (Nobre 2002, p. 19). Since the 1970s and 1980s, when major changes in the Brazilian indigenous policy and constitution occurred, many indigenous groups of mixed descent in the northeast, southeast, and the Amazon region mobilized to attain state recognition of their indigenous status and land demarcation (Oliveira 1999; Warren 2001; Ioris 2005; Mendes 2005; Bolaños 2010). Oliveira (1999) asserts that the ethnic groups of the northeast have gone through a process of ethnogenesis, recuperating presumed extinct indigenous identities and creating new ones. In other words, they have re-emerged as culturally distinct peoples. Hill (1996) asserts that the process of ethnogenesis helps to understand the cultural and political struggles of peoples who want to position their identities and existence within the contexts of historical domination and discontinuity. In Latin America, in the late 20th and early 21st centuries, a wave of emergent peoples has mobilized to position themselves as "historically rooted peoples in contemporary places" (Whitten 2007, p. 376).

After a long process of land struggle, the northeastern indigenous groups were recognized as an ethnic and historical unit, composed of diverse social groups which are culturally and territorially related (Oliveira 1999). Their ethnogenesis process included the re-establishment of indigenous territories, or what Oliveira (1999, p. 20) defines as territorialization, that is, "the process of social re-organization that implies the establishment of a differentiated ethnic identity, the constitution of specialized political mechanisms, redefinition of social control over natural resources, and the re-construction of culture and its relation with the past."

States have also played an important role in processes of territorialization of state power. Regarding the conflicts over the management of the Titicaca National Reserve in Perú, Kent (2008, p. 290) asserts that "appropriation of space by the administrative apparatus has been crucial for the expansion of state control over populations and resources," which has provoked local groups' territorialization as a defensive strategy to state impositions, and have made efforts to regain control over their territories and their way of life. Then, what is at stake in the ethnic and territorial re-definition of the peoples of the lower Tapajós-Arapiuns is the understanding that indigenous territory is not a mere legal or administrative entity to secure the cultural and physical survival.

Indigenous territories are also the result of processes of identity construction. As expressed in the different narratives, the re-construction of indigenous identities constituted a unifying force for the peoples of the lower Tapajós-Arapiuns region to claim their land rights. For instance, a Tapajó indigenous leader explained that before CITA was created, indigenous peoples of the region were both afraid and ashamed of self-identifying as indigenous because of the stigmatization surrounding the term as well as their own personal negative experiences suffered at the hand of such discrimination. As part of their political activism, CITA members created the slogan "I am an Indian and I am not ashamed of it." This slogan not only helped to reaffirm their identity, but also constituted their main political statement. It signifies both Indian pride and belonging to the indigenous movement. In the quote below, a Tapajó indigenous leader from the community of Garimpo explained this sense of pride and belonging:

> Here [in the community] people are happy because we do not feel ashamed anymore of our indigenous tradition ... today, I am not ashamed of asserting my indigenous identity, because I am certain that this was the identity of my ancestors. Today we are clear, we have no doubts ... Today I am what I am, and not what others say. (Interview by author, 12 July 2004)

CITA members' claims to land and ethnic identity rights are framed by both the political struggles to resolve land conflicts, and by the socio-cultural meanings created through historical interaction with a particular territory. Through these claims, CITA members intend to achieve not only the demarcation of their territories, but also the recognition of their ancestral indigenous identities tied to that land. The emphasis on territoriality and land rights as an important component of peoples' ethnicities is always asserted, as one of the indigenous Munduruku from the Takuara community explains:

> We are struggling for something that is ours, our land: the same land that was invaded by the Portuguese a long time ago. We are struggling because we are not afraid anymore. Some people say that we are "turning into Indians" only because we want to secure land. But it is not just a matter of land that makes us assert our identity; it is also a feeling of freedom, the possibility of being what we are inside, to walk our land, to fish in our rivers without being afraid of seeing our land invaded and our people displaced. (Interview by author, 10 June 2004)

However, the territory occupied by CITA members represents a key place for the convergence of environmental and economic interests. There is a global and national interest in establishing plans for the preservation and sustainable management of natural resources, through forest reserves and extractive reserves that have diminished the control of indigenous peoples over their territories, as expressed in the interview with an indigenous Arapium from the community of Vila Franca:

> After the RESEX was created, we discovered that people lost control over their lands ... suddenly there were many organizations deciding what and how to do things. The communities lost their decision-making power, there were governmental organizations such as IBAMA on top of the communities deciding and the communities had just to obey those decisions ... at the beginning we had an association of communities and each had a representative on the administration board of the RESEX, but later that changed and we got only one representative for all communities ... (Interview by author, 8 March 2006)

The new structure established to manage the RESEX requires representatives from government, civil society, as well as each community within the reserve. In reality, however, only one representative from all the communities in the RESEX area actually has a seat on the board. Thus, what "lost control over their lands" refers to in the Arapium leader's previous quote is the subordinate condition of indigenous peoples' interests and management practices to an external agenda. The Arapium leader asserted that by limiting the number of community representatives on the board, not only indigenous peoples but also all other residents of the RESEX area lost decision-making power. Additionally, there is the ongoing process of frontier expansion of soybean production in the region, under the banner of economic development. These two processes have blurred the socio-cultural and political movements of people claiming their territorial and indigenous rights.

Between development and conservation disputes

Santarém is one of the most promising and controversial agricultural frontiers of the country. The rapid expansion of soy farming, together with the paving of the BR-163, is intended to consolidate the export corridor for soybeans via the Amazon River. These projects increased global concerns about rapid deforestation, the effects of global warming, climate change and their related social effects (Goldsmith and Hirsch 2006; Fearnside 2007; Nepstad *et al.* 2008). As soybean cultivation is only profitable at a large scale, displacement and illegal seizure of land has increased (Sauer 2005). In addition, the Cargill soybean export terminal, which started functioning in 2003 without completing the Brazilian environmental licensing process—including the Environmental Impact Study (EIA) and the Report of Environmental Impacts (RIMA)—has complicated these environmental and economic disputes. Various environmental organizations and local governmental agencies protested the establishment of the soybean terminal port pointing out not only its illegal construction, but also its devastating effects in the Santarém region. The international environmental organization Greenpeace published a report titled "Eating up the Amazon," in which Cargill and soy farming were identified as the current major drivers of deforestation (Greenpeace 2006). Since the terminal port opened, several actions by governmental environmental agencies were taken to stop Cargill operations. Finally on 27 February 2007, the Federal Public Ministry (MPF) ordered the closure of the terminal port and requested the fulfillment of the Brazilian environmental license (MPF 2007).

However, the local government, soybean producers, and transporters have viewed the paving and the expansion of soybean farming positively. Their arguments go beyond economic prosperity and employment opportunities to assure that soy farming expansion has no significant environmental threats to the Amazon, because it expands only into savannas and previously converted lands (Brown *et al.* 2005). For instance, in 2004, the Ministry of Agriculture estimated that Santarém had approximately 500,000–600,000 hectares of secondary forest considered to be convertible to agriculture (Escobar 2004). Conversely, environmentalists argued that technological advances not only had allowed soybeans to expand into new climate regions, but had also created pressure to clear and convert primary forest areas to lands used for soy farming, pasture, and secondary forest (Soares-Filho *et al.* 2006).

Some scientists have argued that the environmental and biodiversity value of secondary forest has been undervalued and ignored (Lo-Man-Hung *et al.* 2008). Secondary forests are increasingly dominating the Amazon landscape and scientific studies suggest that secondary forest regeneration can restore conditions suitable to support primary forest species (Lo-Man-Hung *et al.* 2008). This indicates the value of secondary forest for conservation. Secondary forest also constitutes a great source of environmental services, such as watershed and soil protection as well as production of wood and non-timber forest products that benefit many Amazonian peasants and indigenous communities directly.

In the context of these disputes, the indigenous movement experienced difficulties in voicing its claims and gaining support from other organizations. On 1 May 2006, CITA participated in a public demonstration organized by the Frontline Defense of the Amazon (FDA)[9] to protest deforestation, social injustice, the establishment of multinational companies, such as Cargill, and large-scale soy farming in the Santarém region (Santos 2006). This event, which included the participation of Greenpeace, was part of a broad campaign to defend the Amazon forest and forest-dwelling peoples, and was built around the slogan "Impunity kills and deforests the Amazon." At the gathering at Santarém's central plaza, CITA members were confronted by merchants and soy producers who yelled: "Indians are lazy and do not work the land!"; "Land is for those who work it!"; "You are fake Indians, true Indians go naked and live deep in the forest; go back!" (Personal observation, 1 May 2006). Days later, I accompanied CITA to a meeting with the local TV news station-Pontanegra. A conflict had been generated as a result of a TV news broadcast that asserted that the indigenous peoples who gathered at the protest were not real Indians, but rather individuals contracted by Greenpeace to pretend to be Indians (Personal observation, 1 May 2006). The TV news promised to produce a new report to retract the allegation, but never did so.

During the following weeks, several demonstrations both by supporters of the soybean industry and Greenpeace showed an ideological division between conservationists and those in favor of economic development (Agência Amazônica 2006a). In the local newspapers, reports and analyses centered on Greenpeace's radical activism, highlighted the soybean industry's potential to bring economic prosperity to the region, and blurred the FDA protest for social injustice and indigenous peoples' rights claims (Santos 2006a). The local government released a statement highlighting the importance of the soy industry for the economy of Santarém, and called for an inclusive solution of the conflict, in which those in dispute could reach an understanding (Agência Amazônica 2006c). Members of political parties, such as the "Green Party," also released statements condemning the presence of international environmental NGOs in the Amazon region, and defended the soy industry (Santos 2006b).

That same week, The Nature Conservancy (TNC) announced that it had signed the green certification agreement with soy producers. The green certification agreement provides incentives for soy farmers to conform to the Brazilian Forest Code[10] while increasing market prices (Nilder 2006). However, the media response by the environmental organizations focused only on the deforestation caused by the soy industry and the threat to biodiversity, and did not address the land and forest use rights issue of indigenous peoples of the region. Environmentalists did not express concern for the questioning of indigenous peoples identities, nor did they express clear support for the indigenous peoples' land rights claims (Agência Amazônica 2006b). In other words, the environmental conservationist discourse did not help to position the issue of indigenous peoples' tenure rights as an important public matter as CITA members were expecting. Prevailing global emphasis on protecting Amazonian biodiversity and tropical forests diminished the importance placed on preserving the rich cultural and social diversity of the region, and, ultimately, the land rights claims by CITA. In a meeting at CITA's headquarters, indigenous leaders discussed the impacts of the public demonstrations and the involvement of Greenpeace in these events. Although most of the indigenous leaders recognized the environmental NGO as a potentially powerful international ally, they expressed disapproval about how the objectives of the regional social movement for social justice and forest peoples' rights were often blurred by the NGO's own perceived agenda against the multinational company, Cargill (Personal observation 31 May 2006).

A few years later, in November 2009, CITA sent a letter to different mass media organizations to protest the publication of a report by an environmental researcher who intended to question the claims of the indigenous movement asserting that CITA members were "fake Indians" whose only interest was to secure lands (Amazônia 2009; CITA 2009). The research report questioned the indigenous mobilization to defend their lands in the Nova Olinda region, one of the areas of the upper Arapiuns River with major conflicts related to the overlap between private forest concessions, and local and indigenous peoples' territories (Muniz 2006; STTR 2009). Other members of NGOs working in the region have questioned the land rights claims of the communities of Takuara, Bragranca and Marituba by asserting that their founding members were originally from the Arapiuns region (Alloggio 2004). However, this assertion indicates that the NGO members were unaware of the strong kinship network between indigenous peoples from the Tapajós and Arapiuns regions. There have been also complaints about the preliminary studies for land demarcation because members of the FUNAI Technical Group, who developed the preliminary studies for land demarcation seemed to be sympathizers with the indigenous movement (Alloggio 2004).

CITA members are questioned for their presumed "lack of authenticity," and consequently their rights to forestlands are refuted. They do not fit the popular Western romantic images held by many of the colonists in the Santarém area who went there to grow soybeans. For such newcomers, authentic Indians are only those "uncontacted" native peoples who live in primary, intact forest. This prevailing romantic idealism, naturalism, and preoccupation with signs of traditional authenticity conflicts with socio-cultural structures constructed through long and complex processes of change. Fausto and Heckenberger (2007) assert that recent advances in ethnographic, archeological, and historical research in the Amazon have shown that indigenous societies have continuously changed, even before the arrival of Europeans. After 1492, indigenous peoples suffered from the dramatic transformation of their world, such as demographic decimation, mass migration, cultural change, and ethnic revitalization (Fausto and Heckenberger 2007). This led to the formation of new social forms.

Most of the environmental NGOs and governmental institutions have maintained a cautious position in debates regarding the claims of the indigenous peoples of the lower Tapajós-Arapiuns region. The support that the indigenous movement was expecting from these organizations has been delayed until there is legal recognition by FUNAI. This makes the situation difficult for CITA because while their claims to ancestral lands await fulfillment of the state's legal and administrative actions for recognition, they cannot receive decisive support from other organiza-tions. The conflicts related to the overlap between indigenous territories and conservation units can be resolved only when FUNAI finishes the process of land demarcation. Unlike indigenous peoples' international activism and alliances with environmental NGOs during the 1970s and 1980s, which brought substantial benefits regarding land tenure rights and forest protection (Brysk 2000), this example shows that national and international interests suppress and obscure indigenous struggles for tenure and forest use rights.

Zhouri (2004) asserts that international environmental NGOs have fared poorly at minimizing the distance that exists between efforts to maintain biodiversity and improve cultural and socio-historical condi-tions of the Amazonian peoples. Zhouri states, "Amazonia is projected into the global context as a mere ecosystem under the influence of global economic and political forces" (2004, p. 74). The social, cultural, and historical significance that the forest has for indigenous peoples does not have the proper weight in environmental and development discussions. More importantly, these global discussions seem to continue to ignore archaeological and ethnohistorical studies that demonstrate the scale and extent to which Amazonian forests have been modified by indigenous populations (Smith 1980; Gibbons 1990; Coomes 1992; Balée 2003; Heckenberger et al. 2003).

Rather than a pristine or untouched forest, the Amazon region constitutes a constructed landscape in which patterns of biodiversity have been, in large part, shaped by cultural forces (Heckenberger et al. 2007). However, Denevan (1992) asserts that visions of Amazonia continue to be dominated by the "Pristine Myth." This myth posits the idea that, at the time of European invasion, the Americas, and particularly the Amazon, was sparsely populated and occupied by "primitive" egalitarian societies that lived a solitary life in the forest, and subsisted merely through hunting and gathering. It was believed that the Americas was "a world of barely perceptible human disturbance" (Denevan 1992, p. 369). These ideas have reinforced the image of Amazonian indigenous peoples as sparse and isolated social groups that have maintained an unchanging culture for centuries.

The persistence of these ideas has consequences for the current claims of indigenous identity and land rights by CITA members. On the one hand, there is competition for land resources among conservationists, development agents, rural residents, and indigenous peoples. Definitions of forest reserves, extractive reserves, areas for agricultural expansion, and indigenous peoples' lands became difficult to harmonize in the planning of the future environmental and economic development of the region. On the other hand, the dichotomous concept of undisturbed versus disturbed forest has established a parallel dichotomy between "authentic" Indian and indigenous people of mixed descent.

Conservation efforts tend to concentrate on areas of primary forest, mainly because of the persistent idea of "undisturbed conditions," or "lack of human modification," interpreted as the key factor that gives this forest type major biodiversity value. In an analysis of Western environmental policy and education, Gómez-Pompa and Kaus (1992) assert that conservationists have generally conceived the relationship between humans and nature as a dichotomy. In this dichotomy, the forest and wildlife constitute a "wilderness" area enhanced and sustained without people. The authors claim that this "wilderness" conception has led to unrealistic and contradictory views in natural resource management policies (Gómez-Pompa and Kaus 1992).

Although the tropical rainforest constitutes a central topic for articulating the environment at a global scale, the broad differentiation between tropical primary or pristine forest, and secondary forest, has served to apply different conservation and economic values that undermine its importance in bio-cultural conservation. Secondary forest represents the main source of low-cost forest products and environmental services for rural poor communities in tropical countries. However, it does not hold the same economic and biological value as primary forest in development and conservation circles. Rocheleau *et al.* (2001) assert that lands not covered in primary forest are invisible, or illegible, for international forestry and development agencies. For these actors, only large and contiguous areas of primary forest are considered major sources of biodiversity. Rocheleau *et al.* (2001) argue that this idea undermines local agro-forest systems and non-contiguous forest formations that have been shaped by complex processes of historical interaction between people and their environment, and which shelter a variety of tree and wildlife species. According to the ITTO (International Tropical Timber

Organization), secondary forests have the potential to generate significant environmental and livelihood benefits, and "mitigate pressure on primary forest through their ability to produce both wood and non-wood forest products" (ITTO 2002, p. 8).

Governments and development agencies have also appropriated the discourses that differentiate the environmental and economic value of secondary forest to justify the expansion of large-scale cultivation and infrastructural development into the Amazon region. The emphasis on the economic value of nature has established a dichotomy between the biological and market potential of primary and secondary forest. Tragically, this dichotomy affects considerations of the environmental efforts to protect these areas. Because of the relatively low economic potential in terms of availability of goods and environmental services, secondary forest is likely to be replaced by other more profitable resources such as soybean cultivation. The expansion of soy cultivation into areas previously modified or degraded has been used to justify the supposedly low environmental and social impacts of large-scale agro-industry.

Given the belief that the low ecological and economic value of the secondary and disturbed forest provides environmental and moral rights to open new spaces for agricultural development, the claims of indigenous peoples of mixed descent are at risk. Stereotypes about indigenous peoples and their way of life have led to contradictory arguments about the land rights of the indigenous peoples of the lower Tapajós-Arapuns region. On the one hand, they were considered incapable of bringing progress to the region due to their indigenous heritage; therefore, the presence of a new economic actor, the soybean producer, was considered beneficial to the prosperity of the region. On the other hand, when talking about the rights of indigenous peoples, they were considered as lacking authenticity. For many of the newcomers in the Santarém region looking for land to grow soybeans, neither the secondary forest, nor the people inhabiting it, had enough value to be "preserved." Thus, just as the secondary forest can be swept away to establish a more profitable activity, indigenous peoples' land can be taken over to make it productive.

Conclusions

The overlapping of conservation units, frontier agricultural expansion, and indigenous territories in the Amazon region has exacerbated the rights struggles of many indigenous groups who have not yet obtained legal recognition of their lands. These territorial overlaps not only motivated competition for land, but also led to a questioning of the legitimacy of indigenous peoples' claims. Notions of authenticity have become the main political discourse of those opposed to indigenous peoples' rights claims.

In the struggle over land rights, the people of the lower Tapajós-Arapiuns region have re-defined their identity and their landscape as the territory that contains the memory of their indigenous ancestry and culture. Contrary to the state's definition of the lower Tapajos-Arapiuns landscape as a region of conservation units, such as the Flona Tapajos Forest Reserve, the RESEX Tapajos-Arapiuns, and the PAE Lago Grande, the indigenous peoples instead re-defined it as an indigenous territory. Although these types of conservation units are important for biodiversity conservation and the sustainable management of land and forest resources, they are not always accepted by local populations. As shown in the present case, for CITA members the creation of the conservation units in the region contributed in making the process of land claim and demarcation even more complex.

From the reactions to imposed definitions of identity and space used by the state, to collective mobilization to defend their lands from the threats of large-scale agricultural frontier expansion, CITA intends to position the historical and political legitimacy of their rights claims. In the case presented here, the Amazonian landscape has been re-defined by historical struggles over land and forest resources, and more importantly, by local peoples' own processes of identity construction. The configuration of the Amazonian landscape is not only the result of the state regulatory mechanism of forest resources used, international environmental awareness, or development expansion, but also of local peoples' interactions with their land. This case illustrates the different emerging landscapes in contemporary Amazonia that have been re-defined by socio-cultural dynamics and political struggles.

Acknowledgements

I am grateful to the Tropical Conservation and Development Program and Amazon Conservation Leadership Initiative at the University of Florida and the Education For Nature/World Wildlife Fund scholarship that supported my studies and research project in Brazil. I also thank the anonymous reviewers for their insightful comments.

Notes

1. ITTO defines primary forest as a "forest which has never been subject to human disturbance, or has been little affected and that its natural structure, functions and dynamics have not undergone any changes that exceed the elastic capacity of the ecosystem." Secondary forest is "woody vegetation re-growing on land that was largely cleared of its original forest cover" (ITTO 2002, p. 10).
2. The communities are Escrivão, Camarão, Pinhel, Jaca, Jacaré, Paranapixuna, Jauarituba, Santo Amaro, Mirixituba, Muratuba, Limãotuba, Brinco das Moças, Marituba, Alter do Chão, Bragança, Takuara, Vila Franca,

Paricatuba, Braço Grande, Muruci, Nova Vista, Aningalzinho, Amina, Arapiranga, São Miguel, Novo Lugar, São José III, Bom Futuro, São João, Miripixi, Cachoeira do Maró, Arimum, Garimpo, Lago Da Praia, Caruci, Açaizal, and São Pedro.

3. *Caboclo* is a very complex and contested term in Amazonian population literature. People from the region do not refer to themselves as *caboclo*, since this is considered a pejorative term. In what follows, I will not use the term *caboclo* to refer to the people from the lower Tapajos-Arapiuns region, because they do not consider this definition as part of their identity.

4. I have defined "indigenous peoples of mixed descent" as those who, through a historical process of change, have re-defined and re-signified their cultural traditions and social structures. This concept is presented in opposition to the traditional view that considers "mixing" as a destructive force acting against indigenous culture. See Oliveira (1999) for an interesting analysis regarding the notion of "índio misturado," or mixed Indian, Guzman (2006) for the case of Negro river region, and Bolaños (2010) for the case of the lower Tapajós-Arapiuns region.

5. Ethno-political action is understood here as any organized mobilization in pursuit of a group's interests. Brysk (2000) explains that the term refers to the political use of a collective identity and the politicization of cultural practices.

6. According to Asner *et al.* (2005) these five states—Pará, Mato Grosso, Rondonia, Roraima, and Acre—account for about 90% of all deforestation in the Brazilian Amazon.

7. The communities within the area of the extractive reserve are: Escrivão, Camarão, Pinhel, Jaca, Jacare, Parana-puxina, Juarituba, Santo Amaro, Mirixituba, Moratuba, Paricatuba, Brinco das Moças, Vila Franca, São Miguel, Arapiranga, Aminã, Anigalzinho, Nova Vista, São Pedro, and Braço Grande.

8. All direct quotations from interviews were translated by the author from Portuguese.

9. The FDA is a social movement in western Pará, which groups several grassroots organizations, international environmental NGOs, and research and education institutions. CITA has been member of the FDA since 2003.

10. The Forest Code established the legal framework for sustainable use and preservation of forest in Brazil and imposes requirements for sustainable forest management plans. A recent proposal to reform the forest code has created controversy among environmentalists. The proposal intends to re-duce the forest area to be preserved in private lands, which is considered to be an open door to allow agricultural expansion into the Amazon (IPAM 2010).

References

Adams, C., *et al.*, eds., 2009. *Amazon peasant society in a changing environment: political ecology, invisibility and modernity in the rainforest.* London: Springer.

Agência Amazônica, 2006a. Greenpeace x agronegócio: guerra ideologica divide Santarém. *Journal de Santarém e Baixo Amazonas*, May 27–June 2, 7–8.

Agência Amazônica, 2006b. Em nota, Greenpeace manten posicionamento contra soja na Amazonia. *Journal de Santarém e Baixo Amazonas*, May 27–June 2, 7.

Agência Amazônica, 2006c. Prefeita pede paz e diz que ha lugar para todos. *Journal de Santarém e Baixo Amazonas*, May 27–June 2, 8.

Alloggio, T., 2004. Trinta anos da Flona Tapajós: avanços e retrocessos na integração entre conservação ambiental e participação social. *In*: ISA, ed. *Terras indígenas y unidades de conservação da natureza. O desafio das sobreposições*. São Paulo: ISA, 578–585.

Amazônia, 2009. *Pará fabrica índios para ter reservas*. [online]. Available from: http://www.amazonia.org.br/noticias/noticia.cfm?id=334760 [Accessed 13 February 2010].

Appiah-Opoku, S., 2007. Indigenous beliefs and environmental stewardship: a rural Ghana experience. *Journal of Cultural Geography*, 24 (2), 79–98.

Asner, G. P., *et al.*, 2005. Selective logging in the Brazilian Amazon. *Science*, 310, 480–482.

Baleé, W., 2003. Natives views of environment in Amazonia. *In*: S. Helaine, ed. *Nature across culture: views of nature and environment in non-Western cultures*. Boston: Kluwer Academic Publishers, 277–288.

Barreto, F.H., 2009. Traditional peoples: introduction to the political ecology critique of a notion. *In*: C. Adams, R. Murrieta, W. Neves., and M. Harris, eds. *Amazon peasant societies in a changing environment: political ecology, invisibility and modernity in the rainforest*. London: Springer, 95–130.

Bernard, H.R., 2002. *Research methods in anthropology: qualitative and quantitative methods*. New York: Altamira.

Bolaños, O., 2010. Reconstructing indigenous ethnicities: the Arapium and Jaraqui peoples of the lower Amazon, Brazil. *Latin American Research Review*, 45 (3), 63–86.

Brown, C., *et al.*, 2005. Soybean production and conversion of tropical forest in the Amazon: the case of Vilhena, Rondônia. *Ambio*, 34 (6), 462–469.

Brysk, A., 2000. *From tribal village to global village. Indians rights and international relations in Latin America*. Stanford: Stanford University Press.

Charmaz, K., 2006. *Constructing grounded theory: a practical guide through qualitative analysis*. London: SAGE Publications.

CITA, 2006. *Minutes. Ata da reunião realizada em parceria com Resex e movimento indígena*. Santarém.

CITA, 2009. (arapyu@hotmail.com). 25 Nov. 2009. FW: *Carta aberta a população paraense*. Email to Omaira Bolanos (omaira_bolanos@yahoo.com).

Coomes, O., 1992. Blackwater, adaptation, and environmental heterogeneity in Amazonia. *American Anthropologist, New Series*, 94 (3), 698–701.

Creado, E.S., *et al.*, 2008. Between "traditional and moderns:" negotiations of rights in two protected areas of the Brazilian Amazon. *Ambiente & Sociedad*, 11 (2), 225–271.

Denevan, W.M., 1992. The pristine myth. The landscape of the Americas in 1492. *Annals of the Association of American Geographers*, 82 (3), 369–385.

Denzin, N.K. and Lincoln, Y.S., eds., 2003. *The landscape of qualitative research: theories and issues*. Thousand Oaks: Sage Publications.

Escobar, H., 2004. Agricultura em Santarém: progresso ou ameaça? *O Estado de São Paulo*, 1 Feb, p. A12.

Fausto, C. and Heckenberger, M., eds., 2007. *Time and memory in indigenous Amazonia: anthropological perspectives*. Gainesville: University Press of Florida.

Fearnside, P.M., 2007. Brazil's Cuiabá-Santarém (BR-163) highway: the environmental cost of paving a soybean corridor through the Amazon. *Environmental Management*, 39 (5), 601–614.

FUNAI, 2009. *Resumo do relatório circunstanciado de identificação e delimitação da terra indigena Munduruku-Takuara*. FUNAI.

Galvaõ, E., 1979. *Encontro de sociedades: Indios e brancos no Brasil*. Rio de Janeiro: Paz e Terra.

Gibbons, A., 1990. New view of early Amazonia. *Science, New Series*, 248 (4962), 1488–1490.

Goldsmith, P. and Hirsch, R., 2006. The Brazilian soybean complex. *Choices*, 21 (2), 97–103.

Gómez-Pompa, A. and Kaus., A., 1992. Taming the wilderness myth. *Bioscience*, 42 (4), 271–279.

Greenpeace Website, 2006. *Eating up the Amazon*. [online]. Available from: http://www.greenpeace.org.br/amazonia/pdf/cargill.pdf [Accessed 27 November 2010].

Greenpeace Website, 2008. *Amazônia viva: prioridade global. Amazônia*. [online]. Available from: http://www.greenpeace.org/brasil/amazonia/amaz-nia-viva-prioridade-glob [Accessed 25 May 2009].

Guzman, D., 2006. Mixed Indians, caboclos and curibocas: historical analysis of a process of miscegenation; Rio Negro (Brazil), 18th and 19th centuries. *In*: C. Adams, R. Murrieta, W. Neves., and M. Harris, eds. *Amazon peasant societies in a changing environment: political ecology, invisibility and modernity in the rainforest*. London: Springer, 55–68.

Harris, M., 1998. What it means to be Caboclo? Some critical notes on the construction of Amazonian Caboclo society as an anthropological object. *Critique of Anthropology*, 18 (1), 83–95.

Heckenberger, M.J., *et al*., 2003. Amazonia 1492: pristine forest or cultural parkland? *Science*, 301, 1710–1714.

Heckenberger, M.J., *et al*., 2007. The legacy of cultural landscapes in the Brazilian Amazon: implications for biodiversity. *Philosophical Transactions of the Royal Society* B, 362, 197–208.

Hill, D.J., ed., 1996. *History, power and identity: ethnogenesis in the Americas, 1942–1992*. Iowa City: University of Iowa Press.

Ingold, T. and Bradley, R., 1993. The temporality of the landscape. *World Archeology*, 25 (2), 152–174.

Ioris, M.E., 2005. *A forest of disputes: struggles over spaces, resources, and social identities in Amazonia*. Thesis (PhD). University of Florida, Gainesville.

IPAM, 2004. *Ordenamento territorial da BR-163, baixo Amazonas, transamazonica e Xingú. O desenvolvimento que queremos*. Santarém: IPAM.

IPAM, 2010. *ONGs se movimentan para salvar código forestal*. [Online.] Available from: http://www.ipam.org.br/revista/ONGs-se-movimentam-para-salvar-Codigo-Florestal/210 [Accessed 27 November 2010].

ITTO, 2002. *Guidelines for the restoration, management, and rehabilitation of degraded and secondary tropical forest*. ITTO Policy Development Series No. 13.

Kent, M., 2008. The making of customary territories: social change at the intersection of state and indigenous territorial politics on Lake Titicaca, Perú. *Journal of Latin America and Caribbean Anthropology*, 13 (2), 283–310.

Lima, M.L., 2009. *Relatório circunstanciado de identificação e delimitação da terra indígena Cobra Grande (Santarém, Pará)*. FUNAI.

Lo-Man-Hung, N.F., *et al.*, 2008. The value of primary, secondary, and plantation forest for neotropical epigeic arachnids. *The Journal of Arachnology*, 36, 394–401.

Mendes, R.L., 2005. O índio e a questão agrária no Brasil: novas leituras de velhos problemas. *In*: M. Salomon, J.F. Silva, and L. Mendes, eds. *Processo de territorialização: entre a história e a antropologia*. Goiânia: UGC, 11–31.

Merriam, S.B., 2002. *Qualitative research in practice: examples for discussion and analysis*. San Francisco: Jossey-Bass.

Moreira, N., 1988. *Índios da Amazonia: de maioria a menoria*. Rio de Janeiro. Editora Vozes.

MPF, 2007. MPF/PA requisita embargo urgente do porto da Cargill em Santarém. [Online.] Available from: http://www.prpa.mpf.gov.br/2007/mpf-pa-requisita-embargo-urgente-do-porto-da-cargill-em-santarem [Accessed 27 November 2010].

Muniz, A., 2006. Clima de conflito persiste na gleba Nova Olinda. *Jornal de Santarém*. 3–9 June, Atualidades.

Murrieta, S.S. and Winklerprins, A.M.G., 2003. Flowers of water: homegardens and gender roles in a riverine caboclo community in the lower Amazon, Brazil. *Culture & Agriculture*, 25 (1), 35–47.

Nepstad, D.C., *et al.*, 2008. Interactions among Amazon land use, forests and climate: prospects for a near-term forest tipping point. *The Royal Society*, 363, 1737–1746.

Nilder, C., 2006. Municipio de Santarém é o paraíso do econegócio. *Jornal de Santarém eo Baixo Amazonas*, 27 May–2 June, 4.

Nobre, A., 2002. Introdução-reconhecimento das terras indígenas-situação atual. *In*: M.M. Gramkow, ed. *Demarcando terras indígenas: experiencias e desafios de um projeto de parceria*. Brasilia: FUNAI/PPTAL/GTZ, 13–22.

Nugent, S., 1993. *Amazonian caboclo society: an essay in invisibility and peasant economy*. Oxford, UK: Berg Publishers.

Nugent, S. and Harris, M., 2004. *Some other Amazonians: perspectives on modern Amazonia*. London: Institute for the Study of the Americas.

Oliveira, J.P., 1998. *Indigenismo e territorialização: poderes, rotinas, e saberes coloniais no Brasil Contemporâneo*. Rio de Janeiro: Contra Capa.

Oliveira, J.P., ed. 1999. *A Viagem da Volta*. Rio de Janeiro: Contra Capa.

Parker, E.P., 1989. A neglect human resource in Amazonia: The Amazon caboclo. *Advances in Economic Botany*, 7, 249–259.

Perreault, T., 2001. Developing identities: indigenous mobilization, rural livelihoods, and resource access in Ecuadorian Amazon. *Ecumene*, 8 (4), 381–413.

Pontes, F., 2003. *Ata Assambléia de Esclarecimento*. Santarém, Ministerio Público Federal, Procuradoria da Republica, 23 May.

Posey, D. and Baleé, W., 1990. Resource management in Amazonia. *Advances in Economy Botanic*, 7, 1–287.

PPTAL, 2008. (PPTAL@funai.gov.br). 9 April 2008. *Aviso de editais*. Email to Maria Melo (maria.melo@funai.gov.br).

Raffles, H., 2002. *In Amazonia: a natural history*. Princeton: Princeton University Press.

Ramos, A., 2003. The special (or specious?) status of Brazilian indians. *Citizenship Studies*, 7 (4), 401–420.

Rocheleau, D., *et al.*, 2001. Complex communities and emergent ecologies in the regional agroforest of Zambrana-Chacue, Dominican Republic. *Ecumene*, 8 (4), 465–492.

Ross, B.E., 1978. The evolution of the Amazon peasantry. *Journal of Latin American Studies*, 10 (2), 193–218.

Santos, E., 2006. Fracassa público no manifesto. *O Estado do Tapajós*, 2 May, p. 1.

Santos, M., 2006a. Justica de Santarém deferiu liminar contra o Greenpeace. *O Impacto*, 2 June, p. 22.

Santos, M. 2006b. PV critica ativistas e culpa Governo pelos conflitos de terra. *O impacto*, 2 June, p. 22.

Saude e Alegria, 2007. *Uma cartografia da memória. Mapeamento participativo socioambiental. Santarém*[Online.] Available from: htpp:www.saudeealegria.org.br/portal/userfiles/Publicacao_Lago_Grande.pdf [Accessed 16 May 2010].

Sauer, S., 2005. *Violação dos direitos humanos na Amazônia: conflito e violência na fronteira Paraense*. Goiânia: CPT.

Simmons, C.S., *et al.*, 2007. The Amazon land war in the south of Pará. *Annals of the Association of American Geographers*, 97 (3), 567–592.

Smith, N., 1980. Anthrosols and human carrying capacity in Amazonia. *Annals of the Association of American Geographer*, 70 (4), 553–566.

Soares-Filho, B.S., *et al.*, 2006. Modeling conservation in the Amazon basin. *Nature*, 440 (23), 520–523.

Stavenhagen, R., 2005. *Indigenous peoples: an essay on land, territory, autonomy, and self-determination*. Land Research Action Network. [Online.] Available from: http://www.landaction.org/display.php?article = 327, [Accessed 25 May 2009].

Strauss, A. and Corbin, J., 1998. *Basics of qualitative research: techniques and procedures for developing grounded theory*. Thousand Oaks: SAGE Publications.

STTR, 2009. *Nota pública*. Santarém: Sindicato de Trabalhadores e Trabalhadoras Rurais de Santarém.

United Nations, 2007. *United Nations declaration on the rights of Indigenous peoples. Resolution adopted by the General Assembly. Sixty-first session. Agenda item 68*. New York: United Nations.

Vaz, F., 2004. As comunidades Munduruku na Flona Tapajós. *In*: ISA, ed. *Terras indígenas e unidades de conservação da natureza. Os desafios das sobreposições*. São Paulo: ISA, 517–574.

Vaz, F., 2008. *De Caboclos a Indigenas. Uma cronologia da etnogenesi no baixo rio Tapajós* [unpublished paper]. Pará, Brasil.

Vera-Diaz, M.C., *et al.*, 2009. *The environmental impacts of soybean expansion and infrastructural development in Brazil's Amazon Basin. Tufts University.* [Online.] Available from: http://www.ase.tufts.edu.gdae/Pubs/wp/09-05Transport Amazon.pdf [Accessed 5 June 2010].

Veríssimo, A., Lima, E., and Lentini, M., 2002. *Pólos madereiros do estado do Pará*. Belém: Imazon.

Wagley, C., 1976. *Amazon town. A study of the man in the tropics*. Oxford: Oxford University Press.

Warren, J., 2001. *Racial revolutions: antiracism & Indian resurgence in Brazil.* Durham: Duke University Press.

Whitten, E.N., 2007. The *longue durée* of racial fixity and the transformative conjunctures of racial blending. *Journal of Latin American and Caribbean Anthropology,* 12 (2), 356–383.

Zarin, D.J., *et al.*, 2001. Landscape change in the tidal floodplains near the mouth of the Amazon River. *Forest Ecology and Management,* 154, 383–393.

Zhouri, A., 2004. Global-local Amazon politics: conflicting paradigms in the rainforest campaign, Theory. *Culture & Society,* 21 (2), 69–89.

Rubber tapper citizens: emerging places, policies, and shifting rural-urban identities in Acre, Brazil

Jacqueline M. Vadjunec[a], Marianne Schmink[b] and Carlos Valério A. Gomes[c]

[a]Department of Geography, Oklahoma State University, Stillwater, OK, USA;
[b]Center for Latin American Studies, University of Florida, Gainesville, FL, USA;
[c]Secretaria de Estado de Meio Ambiente (SEMA), Acre, Brazil

In the 1970s and 1980s, a strong social movement of rubber tappers in the Amazonian state of Acre achieved remarkable policy goals, gaining new forms of land rights as well as political representation through their alliance with indigenous groups, environmentalists, political parties and human rights advocates. Allies of the social movement entered politics and took state power in 1999 as the "Forest Government," building on the rubber tapper's legacy to embrace the unique cultural and political history of the state, and implementing ambitious plans for forest-based development under the banner of "forest citizenship." In the past 25 years, however, rubber tapper identity has changed rapidly as many rubber tappers migrate to urban areas, or increasingly shift from traditional rubber tapping to more intensive land uses such as commercial agriculture and small-scale animal husbandry. This paper uses data collected from household surveys, key-informant interviews and ethnographic research to explore the idea of what it means to be a "rubber tapper" and "forest citizen" today. We examine the contradictory nature of changing land-use and cultural revitalization efforts among diverse rural and urban populations, and the implications of this diversity for the future of the Forest Government's policies, and the rubber tappers.

Introduction

In the 1970s and 1980s, rubber tappers in the western Amazonian state of Acre achieved the status of environmental icons, and secured remarkable policy goals. Specifically, they gained new forms of land rights as well as political representation, through their alliance with indigenous groups, environmentalists, political parties, and human rights advocates. Close

73

allies of the social movement entered politics and took state power in 1999 as the "Forest Government," building on the rubber tappers' legacy to embrace the unique cultural and political history of the state, and implementing plans for "green" forest-based development to benefit rural communities. In the past 25 years, life for rubber tappers has shifted dramatically as many migrate to urban areas, where they have constructed a city life less strongly rooted in the rubber tapper past. Meanwhile, those who stay in the interior increasingly move away from traditional practices such as non-timber forest product (NTFP) extraction to more intensive land uses such as commercial agriculture and small-scale animal husbandry (Ehringhaus 2005), incorporating urban influences—and even urban second homes—into their evolving lifestyles. As the practices of many rubber tappers have shifted and diversified, their previous identities, rooted in the cultural history of the rubber tapping estates and their successful struggles for land rights, have been reshaped. As a result, rubber tapper and forest identities are increasingly complex and fractured.

The principal achievement of the rubber tappers' movement in the early 1990s was the creation of a novel land reform instrument called the extractive reserve, that grants long-term collective use rights over forested areas to local "extractivist" populations who earn their livelihoods primarily from the extraction of NTFPs from native forests. Since this initial success, Brazilian government agencies have created 67 federal and state extractive reserves in the Brazilian Amazon, spanning nearly 14 million hectares (Gomes 2009). Voters in Acre, elected the Forest Government to three consecutive terms based on its discourses and policies which were reminiscent of the rubber tappers' movement supporting innovative forest development in the state. Moreover, Marina Silva, a former rubber tapper, first served as Brazilian Minister of the Environment from 2003–2008 and, though not elected, successfully brought the green development platform into her electoral campaign for the Brazilian presidency in 2010.

Despite overall support for a continued investment in extractive reserves, green development, and other "working forests"—those intended for sustainable use rather than strict preservation (Zarin *et al.* 2004)—some scholars question the viability of extractivism in terms of its economic, social, and environmental sustainability (Fearnside 1989; Hecht and Cockburn 1990; Murrieta and Rueda 1995). Many argue that non-timber forest extractivism will not alleviate poverty in the long run, and that when a commodity becomes economically important, it is likely to transform into a commercial or large-scale operation (Browder 1990; Homma 1992).

Facing such uncertainties regarding the extractive reserve proposal, Acre's Forest Government undertook ambitious policies to reverse the tendencies towards land-use change, deforestation, and rural-to-urban migration that threatened to undermine the historical basis of the state's economy and society. These proposals were encapsulated in the new term

Florestania, coined to capture the notion of citizenship based in the forest.[1] The state endeavored to recuperate for the state's residents—both rural and urban—a new sense of identity that would celebrate the population's roots in the rubber tapping economy, while breaking from the previous development models to favor environmentally-friendly policies. *Florestania*, was one of the flagships of the new Amazonian "socio-environmental frontier" (B. Becker 2004, pp. 139–158), a uniquely integrated approach to environmental conservation and support for the rights of local peoples that emerged in Brazil in the 1990s from an alliance of environmentalists, local populations and social movements. The Forest Government policies sought to raise the self-esteem of the forest dwelling population, stigmatized because of their poverty and illiteracy, while also drawing urban dwellers into a new *acriano*[2] identity that encompassed social justice and citizenship rights; ethics, transparency and participation in government; pride in Acre and its symbols (Figure 1); conservation of forest resources; and sustainable development—both rural and urban—constructed from the local cultural and ecological history (B. Becker 2004, p. 139; de Sant'Ana Júnior 2004).

While the debate continues surrounding the sustainability of the extractive reserve system and working forests, few published studies have analyzed the rapid changes underway in Acre, in terms of land-use and social identities. Chief among these are widespread shifts in livelihood practices in the extractive reserves, the weakening of the rubber tappers' social movement, and the ascendancy of state-wide sustainable development policies in both rural and urban areas. The historical ties and

Figure 1. The Forest Library in Rio Branco. Note the Acre State Forest Government logo, the outline of a Brazil nut tree, symbolic of *Florestania* (on left). Photograph by Elliott Reid.

current links between rural and urban places and populations have been particularly overlooked, despite their importance for the future of the state's development proposals.

This paper focuses on issues of identity and place to explore the contradictory nature of changing land-use and cultural revitalization efforts in Acre among diverse rural and urban populations, emphasizing their changing practices, political history, and attachment to place. Specifically, we ask: (1) How has the identity of rubber tappers and their social movement evolved with changes in livelihood practices, specifically the decline of rubber in favor of other activities? (2) What new forms of identity have emerged with the Forest Government's sustainable development initiatives in rural and urban areas? (3) How are these changing identities linked to shifting rural and urban landscapes in Acre?

In order to explore these evolving concepts of identity, place, and environmental politics among the iconic rubber tappers of Acre, in the sections that follow we use a hybrid framework based on the role of place and identity construction in cultural geography, as well as environmental identities in political ecology. First, we discuss the research methodology and provide background on the study area. Second, we explore the evolution of rubber tapper identity with respect to livelihood practices, participation in the social movement, and sense of place. Third, we shift scales, focusing on the broader implications of the rubber tappers' movement—the rise of the forest government, the development of forest policy and the emerging concept of "forest citizenship" among urban populations. We conclude by emphasizing the importance of recognizing changing rainforest identities as a realistic basis for future sustainability in the region.

The role of place and identity construction

Much of cultural geography explores the relationship between place, culture, and identity. Tuan (1974, p. 93) calls the human attachment to place, *topophilia*, a "human being's affective ties with the material environment." Places, and the material landscapes they encompass, may have a bio-physical reality but are also constructed and interpreted differently by an individual or group based on past, present and future desires and experiences. A cultural landscape can be seen as "a repository of meanings" based on lived experiences, mundane everyday activities, livelihood practices, societal aspirations, personal attachments, time, and culture, among other things (Tuan 1974, p. 145; Nash 2000; Batterbury 2001). Consequently, landscapes, place, and identities are interrelated. In order to better understand identity, it is argued that landscapes should be read as "text" (Duncan 1990), or increasingly, as multiple, often contested, multi-dimensional, and multi-scalar "textures of place" (Adams *et al.* 2001, p. xiv).

The state of Acre is analyzed in this paper as a socially-constructed landscape whose meaning was forged by migrants and their descendents as they adapted to a new "place" and to social relationships constructed around the extraction of rubber for the world market. This landscape later was embraced and transformed in state-wide policy initiatives of the Forest Government. The emergence of the rubber tapper social movement, its political victories and expansion to other regions through new federal policies, and incorporation as a key symbol of the new state-wide development initiatives, provide rich material for the analysis of the fractured nature of place and identity at different historical moments and distinct geographical scales.

Environmental identities, affinities and social movements in political ecology

Our analysis of the evolution of the rubber tapper social movement into policy is grounded in place theory and engages with diverse themes in political ecology (PE), a subdiscipline emerging from various traditions in both environmental anthropology and human-environment geography that "combines the concerns of ecology and a broadly defined political economy" (Blaikie and Brookfield 1987, p. 17). PE emerged from the growing environmental consciousness of the 1970s, influenced by Marxist critiques (Peet and Watts 1996) of environmental degradation, colonialism, and Third World Development, and post-structuralist notions of multiple, overlapping and often conflicting understandings of environmental and resource management issues, emphasizing the role of structure, power, control, coercion, complementarity and marginality (Schmink and Wood 1992; Rocheleau *et al.* 1996). While various definitions exist, Robbins (2004, p. 5) argues that "political ecology [in all of its expressions] represents an explicit alternative to 'apolitical' ecology."

The PE focus on environmental identity and social movements draws attention to how "changes in environmental management regimes and environmental conditions have created opportunities or imperatives for local groups to secure and represent themselves politically" (Robbins 2004, p. 15). Campbell (1996) shows how women gained space in the rubber tappers' movement merely because increased head counts were needed, but in doing so gained a voice, to some extent, to begin to express women's issues. Social movements that form in response to an environmental issue are often embedded with or used as a platform to express other rights issues. The rubber tappers' movement, while seen by many as an environmental movement, was also at heart a social justice movement, fighting for basic livelihood and territorial rights (Hecht and Cockburn 1990; Allegretti 2002). Bebbington and Batterbury (2001) argue that it is important to look at social movements in terms of differences between and across scales, as well as multiple articulations of such movements. PE seeks to understand these multiple and often conflicting meanings, such as the ways in which state-led sustainable development policies have shaped emerging new identities and the boundaries of forested places in Acre.

A key component to understanding many new social movements involves understanding environmental identity. Haraway (1991, p. 155) argues that identities are complex and "fractured," impossible to define by a single adjective. While many scholars have a long history of romanticizing peasant and indigenous groups (for further discussion, see Bolaños 2011; Turner and Butzer 1992), Haraway (1991, p. 177) tells us that it is time to move beyond long held dualisms such as good/bad, green/not green, nature/culture, and instead see identity as consisting of multiple categories, or affinities. M. Becker (2004) suggests that many Latin American social groups have "hybrid identities" based on an intersecting bricolage of culture and place embedded in their own personal struggles, as well as over-riding social, political, and economic factors. Therefore, a social movement and its collective, constructed, and individual environmental identities will likely mean different things to different social actors over different periods of time.

Methods

This paper draws upon field research carried out by the authors between 2003 and 2005 that explored the idea of what it means to be a "rubber tapper" in Acre today. As part of our fieldwork, Gomes and Vadjunec completed a total of 130 household interviews with rubber tapper families within the Chico Mendes Extractive Reserve (CMER) between March 2004 and January 2005 in conjunction with the Group for Research and Extension in Agroforestry Systems in Acre (PESACRE). Each interview was conducted in Portuguese and, when possible, included both the male and female household head(s). This process took about three hours to complete; we asked both open- and close-ended questions regarding land-use and livelihood activities, as well as open and close-ended questions about rubber tapper identity. Specifically, we asked whether or not the household heads considered themselves to be a "rubber tapper" first and foremost, rather than an agriculturalist, colonist, or small-scale cattle rancher (close-ended question). We then asked the interviewee to define what it meant to be a rubber tapper today (open-ended question). Responses were qualitatively coded using a pile sorting technique.

Additionally, we completed more than twenty semi-structured, open-ended interviews with leaders of the social movement, as well as officials from IBAMA (Brazilian Environmental Agency) at the municipal, state and national levels, municipal heads, state officials, and staff from environmental NGOs with projects in the region. Interviews focused on the institutional actors involved in green development projects in the region, as well as their perceptions of the other actors. These interviews provided context and served to develop a comparison across communities and scales regarding key characteristics of development policy in Acre, as well as an outside understanding of rubber tapper identity. The results of this study are reported in the sections on extractive reserves and rubber tapper identity.

A separate study of the capital city, Rio Branco, tracked changes in the urban population over a 15-year period, from 1989–2004, including the first years of the Forest Government that took power in 1999. Random household surveys of 420 to 800 households were conducted from independent samples of the capital's population drawn every five years (1989, 1994, 1999, and 2004). Teams of local researchers interviewed the self-defined household head, collecting data on all household members as well as the respondent's parents and siblings who were living elsewhere. These data, reported in more detail in Schmink and Cordeiro (2008), provided quantitative information, over time, on the domestic group, economic strategies, migration, perspectives on Rio Branco, life quality, and access to urban services of a representative sample of urban households. That study also shed light on policies to support forestry development and to celebrate the state's history, issues discussed in the sections on the Forest Government and "forest citizenship."

Lastly, we collected ethnographic and other data at various private and public functions in both rural and urban settings. Observing the deployment and consumption of symbolic representations of rubber tapper history, especially differences across urban and rural scales, helped to provide additional insight into the complex nature of rubber tapper and forest citizen identities.

Rubber boom and bust in Acre

Acre is the northwesternmost state in the Brazilian Amazon, bordering Peru and Bolivia (Figure 2). With a current population of just over a half a million, Acre is a small state in demographic terms, representing only 3% of the Legal Amazon's population (IBGE 2005).[3]

Starting with the rubber boom of the late 19th century, the state has had a long history of land conflicts related to territories occupied for the extraction and commercialization of latex from the native Amazonian *Hevea brasiliensis,* or rubber tree, for the world market (Coelho 1982). *Acriano* identity has been intimately and historically enmeshed with the rubber tapper identity that emerged here (Hecht and Cockburn 1990). Tens of thousands of landless poor migrants were encouraged by the Brazilian government to migrate into the rubber territories from the drought-stricken areas of the Brazilian Northeast region, first during the rubber boom at the turn of the 19th century (1870–1914), and later during a brief second rubber boom during World War II (Martinello 1988; Barham and Coomes 1996). In Acre, rubber tappers faced extreme hardships under the

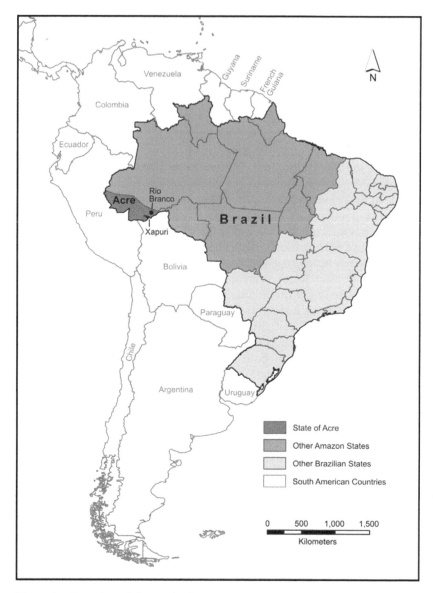

Figure 2. Location of the study site.

traditional debt-peonage (*aviamento*) system dominated by the *patrão,* or rubber barons (de Oliveira Filho 1979). Under this strict system, rubber tappers first began to identify themselves as such, based primarily on the only allowed activity: tapping rubber.

After both rubber booms, many rubber estates were abandoned by the rubber barons, leaving the rubber tappers for the first time free to work the land for themselves. Over several decades, the rubber tappers evolved into an autonomous "forest peasantry" (da Silva 2005), and new generations of Acre-born rubber tappers participated in the construction of a distinctive socioeconomic and cultural institution centered on the *seringal*—encompassing both the rubber groves themselves, and the production and supply system for rubber with its distinctive cultural dimensions (Esteves 1999). Rubber tappers continued to rely heavily on rubber collecting (for local and regional markets) as well as subsistence farming for their livelihoods until the 1970s, when Brazil's military government officially opened up the Amazon for development by outside investors (Santos 1980). In Acre, these policies led to violent land conflicts between rubber tappers and cattle ranchers, where traditional rubber estates were being turned over to cattle ranchers and cleared for the creation of pasture, evicting many rubber tappers and their families in the process (Costa Sobrinho 1992).

The rubber tappers' movement

Starting in the mid-1970s, the rubber tappers began to resist land eviction, with support from the emerging rural worker's unions in Acre at the time (Keck 1995; Allegretti 2002; de Sant'Ana Junior 2004). After a series of violent and deadly conflicts, the rubber tappers began to look for alternative forms of protest and political action, joining forces with local Catholic churches, forming grassroots organizations and encouraging *empates*, or organized, non-violent stand-offs (Bakx 1986; Esteves 1999). The rubber tapper leaders were highly aware of the outside world and crafted their image to respond, in part, to global environmental concerns at the time. They took advantage of the international publicity gained from *empates*, shaping and celebrating their rubber tapper identity—stressing their long and important history in the region, their important place in society as traditional peoples with traditional land uses, their alliance with the indigenous population, as well as their eco-friendly lifestyle. Rubber tapper leader Chico Mendes (1989, p. 72) describes the situation succinctly:

> At the same time as 100 or 200 colleagues are involved in the *empate*, standing in the way of chainsaws and scythes, we aim to have a team whose job it is to get information about what is happening back to Xapuri where another group will make sure it travels all over Brazil and the rest of the world.

In 1985, the National Council of Rubber Tappers (CNS) was created, and the rubber tappers' movement joined forces with various international environmental NGOs pressing for the sustainable development and sustainable livelihoods articulated in the recently published *Our Common Future* (WCED 1987). Rubber tappers came to be viewed by environmentalists as protectors of the forest, extracting mainly rubber and other renewable NTFPs. To ensure the sustainability of both traditional peoples and their forests, rubber tappers called for the immediate creation of extractive reserves, similar to indigenous reserves where, in exchange for land security, they would serve as gatekeepers for vast tracts of forest (CNS 1985). In the *Forest People's Manifesto* published by the CNS, the rubber tappers state:

> We demand a development policy that favors workers. . . . We rubber tappers demand to be recognized as producers of rubber and as the true defenders of the forest. (cited in Hecht and Cockburn 1990, pp. 261–262)

After 1988, the assassination of rubber tapper leader Chico Mendes led to increasing pressure from the international community, and the proposed extractive reserves of the rubber tappers' movement finally became a reality with the creation of the Chico Mendes Extractive Reserve in 1990.

The creation and later expansion of the extractive reserves gave the rubber tappers' movement national and international visibility, and provided the state of Acre with an opportunity to become a leader in carving out a new vision for sustainable development in Amazonia. At the same time, important but less visible changes were underway that provided a rapidly changing scenario for the rubber tappers and other *acrianos*. Shifts in federal and state policies, and in local livelihood systems, led to identity changes throughout the increasingly integrated state-wide rural and urban contexts. While on the one hand, the formalization of the Federal Extractive Reserve System (RESEX) institutionalized a grass roots conservation and development proposal oriented to forest-dwelling peoples, parallel investments in urban industrial development created new conditions that challenged the traditional rubber tapper identity and sense of place, both reinforcing and undermining the forest identities of rural and urban populations in Acre. The impacts of these complex but poorly documented changes on the evolution of practices, places, and identities are explored in the remainder of this paper.

Results

Extractive reserves and rubber tapper identities—evolving practices, places, and identities

The most dramatic change to take place in the *seringal* was the abrupt decline of rubber tapping in the extractive reserves due to federal policy shifts. A reserve-wide study completed immediately following the creation of the CMER showed that rubber was still the main source of income for residents of the reserve (Feitosa 1995, p. 70). Market agriculture played only a minor role in family income, while small animal production and cattle ranching were practically non-existent (Feitosa 1995, p. 70). In the mid 1990s, however, world rubber prices fell, and the federal government suspended protective policies in place for decades that had provided the rubber industry with credits and tax incentives, while regulating the price and import of processed latex.

Residents of the CMER responded by practically abandoning rubber production, although still relying on other NTFPs, such as Brazil nuts, fruits and resins for a portion of their livelihoods, and diversifying their production systems to increasingly include market agriculture, mainly beans, rice, corn and manioc, and small animal production (i.e., chickens, goats, sheep, and cattle). Small-scale cattle ranching is currently the main land-use driving deforestation in the CMER, with some areas of the reserve quickly approaching legal deforestation limits (see Gomes 2009; Vadjunec *et al.* 2009). The "Forest Government" elected to office in 1999 also changed the terrain of place and practice, seeking to reverse land-use changes in the region through a renewed dedication and reinvestment in sustainable forestry, including the (modest) regulation of both rubber and Brazil nut prices in the state of Acre, with the hope of increasing traditional extractivist incomes while halting deforestation (Viana 2004).

These important changes underway in the rural areas were reflected in shifts in the identities of rubber tappers, as revealed by the qualitative coding of open-ended responses to questions on rubber tapper identity from the 130 heads of households surveyed in the CMER in 2003. Results revealed that the great majority of rubber tapper responses (85%) revolved around three main themes: current practice (tapping rubber or not); the historic past and participation in the rubber tappers' movement; and sense of place. The following examples illustrate the ambivalent and shifting views of these dimensions of rubber tapper identity as expressed by most responses gathered in the field.[4]

Rubber tapper practice

The simplest definition of "rubber tapper" involves what one actually does—tap rubber. This is the definition that NGOs, researchers, development agencies and government officials most often use without question. The rubber tappers interviewed, however, expressed this apparently straightforward definition in more nuanced and contradictory ways. Surprisingly, only 33% of households surveyed defined rubber tapper identity based on actual practice. As one resident of the Chico Mendes Extractive Reserve claimed: "You must tap rubber to be a rubber tapper." Yet another rubber tapper explained: "Yes, I still tap rubber. I was born and will die with the name rubber tapper on my lips." Among those who defined rubber tapper identity based mainly on the practice of tapping rubber, 40% surveyed did not actually tap rubber, and therefore did not define themselves as a rubber tapper, but rather as a colonist, agriculturalist, or small-scale cattle rancher. Typical responses include: "No, I am not [a rubber tapper]. I no longer tap rubber" or "I have never tapped rubber in my life."

On the other hand, although less than 42% of the 130 rubber tapper households surveyed tapped rubber, a large majority of these, approximately 78%, still considered themselves to be rubber tappers first and foremost, rather than small-scale agriculturalists, colonists, cattle ranchers or some other designation. Those who continued to tap rubber emphasized their pride in this heritage. The contradictory responses collected in the interviews in the CMER illustrate the complex ways in which changing livelihood practices interacted with other dimensions of rubber tapper identity. For many residents of the CMER the radical shift in their land-use and livelihood practices away from rubber and NTFPs, and towards market agriculture and animal production, came with a real sense of ambivalence and regret. For others, particularly those heavily involved in the movement, there was a growing sense of betrayal. At the annual meeting of the CNS in 2003, one angry rubber tapper accused those residents of the reserve who continued to invest in cattle as literally "stabbing their fallen brother Chico Mendes in the back." Key informant interviews also revealed increasing tension between NGOs, government agencies, and rubber tappers, with some agencies perceiving rubber tappers as having "abandoned their class." As one local municipal leader explained, "The rubber tapper does not exist anymore. He is dead." When asked the question of what it means to be a rubber tapper today, one household head took off his hat, shook his head and said, "The reserve today is full of colonists. No one is a rubber tapper anymore."

While some respondents glorified the bygone past, others in the reserve remarked on the low status of rubber tappers in Acre. Respondents equated "rubber tappers" with poverty, lack of access to infrastructure, education, and healthcare. As one rubber tapper aptly put it, "to be a rubber tapper is to belong to a class that everyone wants to forget."

Historic past and participation in the movement

Rubber tappers in the CMER also frequently referred to the history of the rubber tappers' social movement as a key dimension of rubber tapper identity. The movement was initially successful because of both its visibility and the high level of participation in it of men, women, and even children. Although local leaders often claimed in key informant interviews that participation had fallen off, household surveys revealed that approximately 50% of households still actively participated in the community and municipal—level rubber tapper associations and rural worker's unions. Many rubber tappers (25%) were united by and continued to define themselves primarily based on their historic past ties to rubber and/or their participation in the movement. As one rubber tapper argued, "even though we do not tap rubber anymore, we are united together in our fight as rubber tappers. We stood beside Chico Mendes to win this reserve."

Many expressed their historic ties to tapping rubber, even if not necessarily actively associated with the movement itself. Respondents celebrated, respected, sometimes even romanticized past traditions of rubber tapping, as one rubber tapper explained, "I do not extract rubber anymore, but I am still a rubber tapper. I was created on the milk of the rubber tree. We all were. She created us."

For some, the name "rubber tapper" was seen as a great source of "pride" or "value." Others, however, lamented that "rubber tapper" was beginning to lose its meaning because of the disjuncture between past and present practices.

Rubber tapping was also an activity to which some residents hoped to return should the price of latex improve enough to make it worthwhile. As one resident told us emphatically, "we all have a history of tapping rubber, but many of us do not tap rubber anymore because the price is bad." The strength of the past heritage, and the hope that current changes might be transitory, combined to favor the persistence of rubber tapper identities.

Place

While rubber tappers had increasingly complex, multi-use livelihood trajectories, more than one-quarter began to define themselves more by *where* they lived than *what* they did. With the rich tapestry of long, interwoven, and continuously evolving histories surrounding the importance of rubber in Acre, the success of the social movement with the creation of the CMER, and the election of and support provided by the Forest Government, people had become more deeply united by a place, a philosophy, and the forest. Overall, 28% of respondents defined being a rubber tapper by where they lived, not by what they did or their past histories.

The symbol of working forests had expanded; it now represented not only the reserve, but also Acre, as a larger all-encompassing landscape linked to the practices associated with the forest as well as the place of the forest. For example, a household head who defined himself as a rubber tapper, even though he had not tapped rubber in many years, explained how the past practice of tapping rubber was intertwined with the struggles of the social movement to win their "place" in the reserve:

I grew up tapping rubber—for this reason, I am a rubber tapper. Without the rubber tapper, we would not have this reserve today. We would not have this reserve without our name, our blood, our sweat. For these reasons we are all rubber tappers.

For many, being a rubber tapper meant much more than just tapping rubber; instead it meant "living in the reserve," residing in a "special place," being a "guardian of the reserve," or "caring for the forest." As one rubber tapper explained while pointing to the surrounding forest, "Everyone who lives *here* is a rubber tapper." This strong sense of place— a place defined as the forest—was the basis for the Forest Government's proposals for sustainable development with "forest citizenship." Being a rubber tapper even had the ability to transcend the reserve, as well as the rural-urban divide, albeit imperfectly with the rapid changes underway. As one rubber tapper explained:

> Rubber tappers aren't just those who tap rubber. Whoever lives in the *seringal* is a rubber tapper. Whoever lives in the reserve is a rubber tapper. In the city there are even rubber tappers. Sadly, the name "rubber tapper" has been prejudiced. It's a dirty word.

Residents of the CMER often commented on the pejorative connotation of rubber tapper identity because of its association with illiteracy and poverty. For others, rubber tapper identity had become negative more recently, mainly because it was sometimes used by NGOs and government agencies as a means of controlling the definition of acceptable and un-acceptable land-use and livelihood activities in the reserves, often against the aspirations of rubber tappers who sought to diversify their livelihood practices and increase their cattle herds.

Not only were residents of the CMER increasingly involved in mixed land-use activities, but they were also increasingly urbanized. Their children often had no option but to attend middle and secondary school in the nearest outside cities. The majority of the reserve's residents had relatives who lived in the surrounding urban areas, and 30% of the CMER's current residents came either from a colonization project, urban, or peri-urban area before moving into the CMER. In 2003, 10% of the rubber tappers surveyed already owned a house in the city, and approximately 10% were thinking of moving outside the reserve to the surrounding urban areas.

The surprising persistence of the rubber tapper identity, even with the sharp decline of rubber tapping as a key livelihood practice over more than a decade, the turnover in population, and increasing ties to urban areas, reveals the complex and contradictory nature of rubber tapper identity. Deep roots in cultural history, in the experience of political mobilization in the rubber tapper social movement, and the broader consciousness of their role as "forest guardians," as well as an attachment to the reserve as a place for which the movement had fought hard, kept alive the rubber tapper social and political identity even in the face of rapid livelihood changes. These contradictory aspects of identity (being a "rubber tapper") versus practice (no longer tapping rubber) emerged, in part, in response to the sustainable development and cultural revitalization policies of the Forest Government that was elected to office in 1999, and whose political project specifically was grounded in the history of rubber tappers, their struggles, and their place—the forest.

.

The Forest Government and "forest citizenship"—urban-industrial identities and places

Translating the lofty goals of "forest citizenship," or *Florestania*, into practice required the Forest Government to adopt a series of strategies to reverse changes in land-use practice, affirm the historical and cultural roots of the extractivist past, and forge a new sense of place among *acrianos* that could encompass both rural and urban areas, while investing heavily in the modernization of the state's economy. The resulting policies had important impacts in the complex and evolving identities of both rural and urban populations of the state.

Forest development policies

One of the first policies implemented in 1999 was the Chico Mendes law, which was intended to revive the practice of rubber tapping, as well as NTFP extraction, the mainstays of the traditional rubber tapper identity. The measure provided an additional payment per kilo of rubber to tappers who had their documents and were members of legally-constituted producer associations (Kainer *et al.* 2003). Reversing the federal government's withdrawal of rubber subsidies in the 1990s, the state measure provided an important immediate benefit to its political base among rubber tappers, endeavoring to keep the extractivists living in the forests and to promote their organization as producers and as citizens. The government also created new institutions to focus on forest production and marketing, including a State Forestry Secretariat and a network of regional cooperatives linked to a state-level organization, COOPERACRE. Ten years later, this rubber-support policy would bear fruit with the installation of a state-of-the-art condom factory managed by the state government. This factory buys native rubber from hundreds of rubber tappers and makes it into condoms that are sold to the federal government's Department of Health. Brazil nut processing factories were also established in Xapurí and in Brasiléia. The government sought to subsidize the continuation of the practice of tapping rubber and support the development of complementary NTFPs such as Brazil nuts, *copaiba*, and *andiroba* oils. As rubber tappers in different regions began to involve themselves in these new initiatives, their identity broadened from that of "rubber tapper" to "extractivists"—a new label later adopted by the National Rubber Tappers Council to acknowledge its changing constituency. For those who opted to remain in the extractive reserve instead of migrating to urban areas, these policies provided an important stimulus to revive their extractivist identities and ties to the forest, now in a more modern guise.

Alongside these investments in extractivist product marketing to strengthen the identity-based practices of former rubber tappers, the state also made controversial major investments in sustainable timber management (Kainer *et al.* 2003). Given the difficulties and uncertainties of developing diversified NTFP markets, the certain profits to be had from the growing demand for tropical timber were seen as a key component of the state's development strategy. These investments were funded by the state, as well as major loans from the National Economic and Social Development Bank (BNDES), and the Inter-American Development Bank. The state's ambitious forest policies focused on both commercial and community forest management, and were supported by a new Forestry College.

In order to develop urban-based value-added processing of forest goods from rural areas, the state stimulated the creation of Furniture Development Growth Poles in both Rio Branco and Xapuri, and brought in outside consultants for the design and development of high quality projects for both internal and external markets. These investments were especially important in Xapuri, the historic heart of Acre's rubber tapper cultural history and the site of Chico Mendes' home, where he was assassinated in 1988. These important events in Acre's cultural history were celebrated with the construction and refurbishing of local museums and an incipient tourism initiative in Xapuri, alongside the installation of a high-tech flooring factory and a dozen other forest industries.

By 2008, the forest sector had increased its share from 7.4% to 18% of the state's production, and it accounted for half of the state's exports; the proportion of processed wood products coming from managed forests grew from only 5.7% in 2002 to 84% in 2008 (Acre State Government 2008). The Antimary State Forest became the first and only certified public forest in Brazil. The state's investments in public forests and public-private joint ventures paid off primarily in the expansion and modernization of commercial logging enterprises. Among extractivists, the response was more ambivalent. Technical difficulties and political opposition slowed the state's plans for implementation of community forest management in dozens of extractivist communities. The lingering strength of the "rubber tapper" or "extractivist" identity among many forest residents, and the memory of the historical struggle against outside loggers and ranchers, led many rubber tappers to resist transforming their "place" into a site for timber extraction. On the other hand, some former rubber tappers embraced timber management as a viable strategy to modernize and diversify their forest production systems, while still struggling with the uncertainties of techniques for sustainable timber management, including the costly and stringent requirements for timber certification.

Forest citizenship

In addition to policies promoting the revival of extractivist practices, the state also invested in celebrating the state's unique history rooted in the forest, seeking to improve the self-esteem of rural inhabitants previously dismissed as invisible, poor, and backward. In the state capital, Rio Branco, the Forest Government's initial investments focused on revitalizing the historical center of the city (palaces, museums, libraries, parks, riverside, markets) that celebrated the state's strong sense of autonomy from the federal government,[5] as well as the rubber tapper history and political mobilization, symbolized by the statues of Chico Mendes, the Chico Mendes Park and Zoo, a modern stadium called the Forest Arena, a modern public library called the Forest Library, and the remodeled historic commercial center dating from the rubber boom at the turn of the century. (See Figures 1, 3 and 4.) The Chico Mendes prize, awarded each year on the anniversary of the rubber tapper leader's death, was created to recognize people and programs that contribute to sustainable development in the state.

In 2009, the state government launched a major program called "Digital Forest" with the aim of providing free wireless internet access for cities throughout the entire state (Figure 5). Implemented first in Rio Branco, it made this city the first Brazilian capital city with full free internet access. A public radio station ended each nightly broadcast with the state's anthem, strengthening residents' sense of pride as *acrianos*. The state government also sponsored a new historical opera, and a nationally-telecast mini-series, *Amazônia, De Galvez a Chico Mendes* (TV Globo

Figure 3.   A symbol of Acre's illustrious past: the recently renovated Governor's Palace, originally built in 1930, Rio Branco, Acre. Photograph by Carlos Valério A. Gomes.

Figure 4. Statues of Chico Mendes and son, Sandino, at Rubber Tappers' Square in downtown Rio Branco. Photograph by Carlos Valério A. Gomes.

2007), that brought Acre's rich cultural heritage to a broader audience as well as to residents of the state capital. Rubber tapper poetry, artwork and music regularly kicked off public events.

These multiple initiatives sought to transform the state's history from a heritage of isolation, poverty and backwardness, to one of heroism and unique pride. Alongside efforts to build identity through supporting extractivist practices and to celebrate the state's history, the government also sought to reinforce pride of place by forging a state-wide identity as *acrianos*. A new form of citizenship—*Florestania*—would encompass

Figure 5. Rio Branco's urban "Digital Forest" with free Wi-Fi access. Photograph by Carlos Valério A. Gomes.

the rural cultural roots in the forest but also deliver the benefits of development that previously had been denied to all but the state's elites. Residents of Rio Branco enjoyed improvements such as the construction of principal thoroughfares and recuperation or construction of new access highways (the Chico Mendes Highway), including a new airport and new bridges, as well as growth of the industrial sector (Moreira 2006). (See Figures 6 and 7.)

Urban *Florestania*

With these public investments, Rio Branco's "urban forest" was becoming the command center for the sustainable development project of the Forest Government in Acre (Becker 2005). The Forest Government's policies led to growth of Acre's gross domestic product at rates higher than those for Brazil as a whole, and for the Amazon region, from 1989 to 2003 (IBGE 2003, cited in Schmink and Cordeiro 2008, p. 99). The social benefits from these policies were reflected in the improved life expectancy, reduced infant mortality rates, increased literacy and higher education levels of *acrianos*, especially those living in the capital.

An accelerated process of rural-to-urban migration in Acre, beginning in the 1970s, led to an increase of the urban population (as defined by the Brazilian census bureau) by over 500% in 2000, growing from 28% to 66% of the total population (IBGE 2005). These migrants sought out Rio Branco in large part due to the historically unequal process of economic development, which generated important differences in life quality

Figure 6. Recently renovated trading and counting houses celebrating Acre's rubber tapping past ("the old market") meets bustling downtown Rio Branco. State of the art pedestrian suspension bridge takes forest citizens back and forth across the Rio Acre bridging the old and new business districts. Photograph by Carlos Valério A. Gomes.

Figure 7. A new road built under the Forest Government (with Forest Government logo/Brazil nut tree) serves Rio Branco's new international airport and growing industrial zone. Photograph by Carlos Valério A. Gomes.

between rural and urban areas, and between the capital and other cities and towns. Strong family ties also influenced the migration process, and are cited even more often than economic reasons for moving to the capital. Most of the people living in Rio Branco had previously lived in a rubber tapping area, revealing the still-strong rural roots in the urban setting. However, their previous *urban* experience was far stronger: a large and growing percentage had lived in another city before moving to Rio Branco (73% in 1980, growing to 93% in 2004). In 2004, approximately 72% migrated directly to Rio Branco from another urban area. Data on employment and migration patterns of the current and previous generation of Rio Branco's population reveal a shift into urban, salaried employment from rural informal or self-employment. The urban population was distancing itself from its forest extractivist past, while at the same time receiving many of the benefits of *Florestania*.

These results reveal a profile of predominantly internal interurban migration, no longer a flow dominated by rubber tapper refugees. Moreover, approximately 60% of Rio Branco migrants said they would not return under any conditions to the rural area, and 80% were not interested in returning to rubber tapping areas. In response to government policies to improve conditions there, more people (30% in 2004 compared to only 16% in 1989) began to express interest in returning to the rural areas if they received land. This incentive appealed primarily to less wealthy urban residents. Among the poorest half of the Rio Branco population interviewed, slightly more than half said they would consider returning to the rural areas if they received land—twice as likely as the wealthy residents interviewed.

Although differences persisted between the richest and poorest families in Rio Branco, significant improvements emerged in many social indicators for all urban social groups under the Forest Government. Compared to other rural and urban areas of Acre, Rio Branco's residents enjoyed more employment opportunities, better-quality housing, increased access to cooking fuel and other goods, and to urban services including electricity, garbage collection, piped water, access to public transport, churches, schools, health posts, leisure facilities, and local commercial enterprises such as groceries, butcher shops, bakeries, pharmacies, and other stores—which multiplied rapidly after 1999 (Schmink and Cordeiro 2008). Residents of the capital (especially poorer residents) were increasingly hopeful about the future impact of the city. The tangible improvements in life quality, and the efforts of the Forest Government to strengthen pride in Acre's culture and history, contributed to the *acriano* sense of place under the banner of *Florestania*, and encouraged most urban dwellers to want to remain in the capital, despite the limitations and inequities of the economy of the capital. While the long-term sustainability and environmental impact of *Florestania* policy remains uncertain, the Forest Government's development policy remains a remarkably ambitious attempt to improve quality of life, well-being, and confidence of its citizens to make Acre "the best place to live in the Amazon region" (Acre State Government 2008, p. 14).

Emerging forest citizens: bridging the rural-urban divide

The rapid socio-economic, cultural and political changes among rural and urban populations of Acre discussed here reveal the fluid and fractured nature of the identities being forged under the influence of both the rubber tappers' movement and *Florestania* policies. In the extractive reserve named for the rubber tappers' famous leader, residents struggled to reconcile the heritage of rubber tapping practices, successful social movement struggles, and a deep pride in the forest as their place, with the persistent poverty associated with a rubber economy in decline, the regret and anger over increasing deforestation in the reserve, and resentment of the imposition of outsider definitions of rubber tapper livelihood practices. The rubber tapper identity persisted, with its roots in practice, history, and place, despite the rapid changes underway in their livelihood systems, social movements, and the forest itself. Among social movement leaders, the adoption of the broader identity of "extractivists" encouraged by statewide initiatives served as a bridge to this new reality, maintaining the essence of the forest steward identity. Especially among younger or more recent residents, however, many had abandoned the pejorative identity of rubber tapper in favor of identities associated with the more modern practices of agricultural and cattle production, and living in urban areas.

In the capital city of Rio Branco, urban migrants and growing urban-born generations commingled as families combined rural and urban residences, and the important political and cultural history of the rubber tappers was both reinforced by the Forest Government's policies, and undermined by the urbanizing and modernizing development processes set in motion in the ambitious state development plans. With a substantial majority of the state's population living in urban areas, the concept of "forest citizen" was crafted as a new all-encompassing identity associated with the emerging identity of the state, still tied to the past practices and places that made it unique. As a citizenship project, *Florestania* policies were intended to both counterbalance the inherently unequal impacts of development (in the classic sense of citizenship; see Marshall 1950), and to forge a strong state identity that could unite people across generations and bridge the rural-urban divide.

This research reveals the tensions and fluid relations among urban and rural forest citizens in Acre. While rubber tappers continued to be celebrated in the city as part of the discourse of *Florestania*, through music, opera and poetry among other things, rubber tappers regularly complained about the government's neglect of the rural areas. When asked about the benefits of *Florestania*, rubber tappers acknowledged the better prices received for NTFP products, and other advances, but some also argued that too few of the benefits of the Forest Government actually reached the rubber tapping areas; rather, they perceived the bulk of benefits to remain in Rio Branco, far away from the actual forest. One rural municipal official declared, "Where is the Forest Government? We are here still waiting. . . ." Such comments reveal the limits of the reach of the Forest Government's development plans.

Conclusions

This study has focused on how the socially-constructed landscape that is now the state of Acre has evolved over more than a century of migration, influenced by rubber boom and bust cycles, as well as more recent liveli-hood changes, and political mobilization. Rapid changes underway in Acre since the initial victories of the rubber tappers in the 1990s have gone relatively unnoticed in the literature on this well-known social movement, although this research reveals that they have set in motion contradictory shifts in identity and sense of place as other livelihood practices have increasingly taken precedence over tapping rubber in the extractive reserve. Instead, new and expanded identities ("extractivists") have emerged that strategically broaden the social movement identity while maintaining the crucial tie to the forest as their place, and new official brands ("forest citizens") embrace the rubber tappers and their forest as part of the statewide identity, and tie the forest to a modernizing project whose effects, contradictorily, often benefit the more wealthy urban dwellers over the still-isolated and poor forest dwellers.

As new generations take their place in both rural and urban spaces in Acre, the distance among and between the many forest identities may become ever greater. The complex nature of evolving identities and lifestyles in the state provide a significant challenge for realistic development policies that can address multiple and sometimes contradictory demands. By celebrating forest lifestyles, *Florestania* carved out a much-needed, legitimate space for forest people as they became part of the increasingly dominant urban Amazonian landscape. As more people move to urban areas, *Florestania* serves to provide an important link to the forest through the idea of forest citizenship, while at the same time forcing the government to invest more heavily in urban infrastructure, thus, to some extent, encouraging even greater urbanization. *Florestania* will continue to be successful to the extent that both rural and urban peoples achieve equal voice and benefits in the development process, in accordance with their changing identities, needs, and aspirations.

Acknowledgements

This research was funded, in part, by a Fulbright-Hays Doctoral Dissertation Research Abroad Fellowship from the U.S. Department of State, the Program de Doutorado Pleno no Exterior/CNPq, and Programa BECA/Instituto Internacional de Educação do Brasil. We are grateful to Mâncio Lima Cordeiro, and the Group for Research and Extension in Agroforestry Systems in Acre (PESACRE) for continued support. We would also like to thank Mike Larson for his cartographic expertise, and Alyson Greiner, Jay H. Jump, Rebecca Sheehan, and two anonymous reviewers for many helpful comments regarding this manuscript.

Notes

1. The term was coined by Antônio Alves Leitão Neto, an advisor to the state government in Acre.
2. People from Acre, according to the new spelling rules instituted in Brazil in 2008.
3. The "Legal Amazon" refers to the official designation of the Brazilian Amazon Region according to the Brazilian Institute of Geography and Statistics (IBGE). This political and administrative region has expanded over time (see Browder and Godfrey 1997, p. 18), to include the states of Acre, Amapá, Amazonas, Mato Grosso, Maranhão, Pará, Rondônia, Roraima, and Tocantins.
4. Given the sensitive nature of the topic and for ease of reading, direct quotations are presented here anonymously, and were translated by the authors from Portuguese. Quotations were selected from household interviews completed with rubber tapper families within the Chico Mendes Extractive Reserve (CMER) between March 2004 and January 2005 (see methods section for further detail).
5. The state briefly was an independent country at the turn of the 20th century, from 1899 to 1900 (see Schmink and Cordeiro 2009).

References

Acre, State Government Website, 2008. *Plano Plurianual 2008/2011*. Available from: http://www.ac.gov.br/index.php?option = com_docman&task = cat_view &gid = 27&Itemid = 99999999 [Accessed 28 July 2010].

Adams, P.C., Hoelscher, S., and Till, K.E., 2001. Place in context: rethinking humanist geographies. *In*: P.C. Adams, S. Hoelscher, and K.E. Till, eds. *Textures of place: exploring humanist geographies*. Minneapolis: University of Minnesota Press, xiii–xxxiii.

Allegretti, M., 2002. *A construção social de políticas ambientais: Chico Mendes e o movimento dos seringueiros*. 2002. Thesis (PhD). Universidade de Brasília.

Bakx, K.S., 1986. *Peasant formation and capitalist development: the case of Acre, Southwest Amazonia*. Thesis (PhD). University of Liverpool.

Barham, B.L. and Coomes, O.T., 1996. *Prosperity's promise: the Amazon rubber boom and distorted economic development*. Oxford: Westview Press.

Batterbury, S., 2001. Landscapes of diversity: a local political ecology of livelihood diversification in South-western Niger. *Ecumene*, 8 (4), 437–464.

Bebbington, A.J. and Batterbury, S., 2001. Transnational livelihood and land-scapes: political ecologies of globalization. *Ecumene*, 8 (4), 369–380.

Becker, B., 2004. *Amazônia: Geopolítica na virada do III milênio*. Rio de Janeiro: Garamond.

Becker, B., 2005. Amazônia: nova geografia, nova política regional e nova escala de ação. *In*: M. Coy and G. Kohlhepp, eds. *Amazônia Sustentável: Desenvolvimento sustentável entre políticas públicas, estratégias inovadores e experiências locias*. Rio de Janeiro: Garamond, 23–44.

Becker, M., 2004. Peasant identity, worker identity: multiple modes of rural consciousness in highland Ecuador. *Estudios Interdisciplinarios de America Latina y el Caribe*, 15, 1–26.

Blaikie, P. and Brookfield, H.C., 1987. *Land Degradation and Society*. London: Methuen.

Bolaños, O., Forthcoming 2011. Redefining identities, redefining landscapes: indigenous identity and land rights struggles in the Brazilian Amazon. *Journal of Cultural Geography*.

Browder, J.O., 1990. Extractive reserves will not save the tropics. *BioScience*, 40, 626.

Browder, J.D. and Godfrey, B.J., 1997. *Rainforest cities: urbanization, development, and globalization of the Brazilian Amazon*. New York: Columbia University Press.

Campbell, C., 1996. Out on the front lines but still struggling for a voice: women in the rubber tappers' defense of the forest in Xapuri, Acre, Brazil. *In*: D. Rocheleau, B. Thomas-Slayter, and E. Wangari, eds. *Feminist political ecology: global issues and local experiences*. London: Routledge, 27–61.

CNS, 1985. Forest peoples' manifesto. First Amazonian rubber tappers meeting, 11–17 October 1985 Brasília, Brazil.

Coêlho, E.M., 1982. *Acre: o ciclo da borracha (1903–1945)*. Dissertation (PhD). Universidade Federal Fluminense (Niteroi, Brazil).

Costa Sobrinho, P.V., 1992. *Capital e trabalho na Amazônia occidental*. São Paulo: Cortez.

da Silva, S.S., 2005. *Resistência camponesa e desenvolvimento agrário na Amazônia-Acriana*. Dissertation (PhD). Universidade Estadual Paulista.

de Oliveira Filho, J.D., 1979. O caboclo e o brabo: notas sobre duas modalidades de força-de-trabalho na expansão da fronteira amazônica no século XIX. *Encontros com a Civilização Brasileira*, 11, 101–140.

de Sant'Ana Jr., H.A., 2004. *Florestania: A saga Acreana e os povos da floresta*. Rio Branco: Universidade Federal do Acre.

Duncan, J., 1990. *City as text: the politics of landscape interpretation in Kandyan Kingdom*. Cambridge: Cambridge University Press.

Ehringhaus, C., 2005. *Post-victory dilemmas: land-use, development, and social movements in Amazonian extractive reserves (Brazil)*. Dissertation (PhD). Yale University.

Esteves, B.M.G., 1999. *Do "manso" ao guardião da floresta*. Dissertation (PhD). Universidade Federal do Rio de Janeiro.

Fearnside, P.M., 1989. Extractive reserves in the Brazilian Amazonia. *BioScience*, 39, 387–393.

Feitosa, M.L., 1995. Chico Mendes Extractive Reserve. *In*: M.R. Murrieta and R.P. Rueda, eds. *Extractive reserves*. Cambridge: IUCN, 69–76.

Gomes, C.V.A., 2009. *Twenty years after Chico Mendes: extractive reserves' expansion, cattle adoption and evolving self-definition among rubber tappers in the Brazilian Amazon*. Dissertation (PhD). University of Florida.

Haraway, D.J., 1991. *Simians, cyborgs, and women: the reinvention of nature*. New York: Routledge.

Hecht, S. and Cockburn, A., 1990. *Fate of the forest: developers, destroyers and defenders of the Amazon*. Verso: Routledge.

Homma, A.K.O., 1992. The dynamics of extraction in Amazonia: a historical perspective. *In*: D. Nepstad and S. Schwartzman, eds. *Non-timber products from tropical forests: evaluation of a conservation and development strategy*. Bronx, N.Y.: The New York Botanical Garden, 23–91.

Instituto Brasileiro de Geografia e Estatística (IBGE), 2005. *Pesquisa nacional por amostra de domicílios*. Rio de Janeiro: IBGE.

Kainer, K.A., *et al.*, 2003. Experiments in forest-based development in Western Amazonia. *Society and Natural Resources*, 16, 869–886.

Keck, M., 1995. Social equity and environmental politics in Brazil: lessons from the rubber tappers of Acre. *Comparative Politics*, 27 (4), 409–424.

Marshall, T.H., 1950. *Citizenship and social class, and other essays*. Cambridge: Cambridge University Press.

Martinello, P., 1988. *A "batalha da borracha" na segunda guerra mundial e suas consequências para o Vale Amazônico*. Rio Branco: Federal University of Acre.

Mendes, F.A., 1989. *Fight for the forest: Chico Mendes in his own words*. London: Latin American Bureau.

Moreira, R., 2006. Acre começa a entrar na era da industrialização. *Página 20*, 17 July, pp. 12–13.

Murrieta, J.R. and Rueda, R.P., 1995. *Extractive reserves*. Cambridge: IUCN.

Nash, C., 2000. Performativity in practice: some recent work in cultural geography. *Progress in Human Geography*, 24 (4), 653–664.

Peet, R. and Watts, M., 1996. *Liberation ecologies*. London: Routledge.

Robbins, P., 2004. *Political Ecology: A Critical Introduction.* Oxford: Blackwell.

Rocheleau, D., Thomas-Slayter, B., and Wangari, E., 1996. Gender and environment: a feminist political ecology perspective. *In*: D. Rocheleau, B. Thomas-Slayter, and E. Wangari, eds. *Feminist political ecology: global issues and local experiences.* London: Routledge, 3–22.

Santos, R., 1980. *História econômica da Amazônia.* Belém: IDESP.

Schmink, M. and Cordeiro, M.L., 2008. *Rio Branco: cidade da florestania.* Belém: Universidade Federal do Pará/Universidade Federal do Acre.

Schmink, M. and Wood, C.H., 1992. *Contested frontiers in Amazonia.* New York: Columbia University Press.

Tuan, Y-F., 1974. *Topophilia: a study of environmental perception, attitudes, and values.* New York: Columbia University Press.

Turner II, B.L. and Butzer, K.W., 1992. The Columbian encounter and land-use change. *Environment*, 34, 16–20.

TV Globo. 2007. *Amazônia, De Galvez a Chico Mendes.* Globo Comunicação.

Vadjunec, J.M., Gomes, C.V.A., and Ludewigs, T., 2009. Land-use/cover-change among rubber tappers in the Chico Mendes Extractive Reserve, Acre, Brazil. *Journal of Land Use Science*, 4 (4), 249–274.

Viana, J., 2004. Foreword. *In*: D.J. Zarin, *et al.*, eds. *Working forests in the neotropics: conservation through sustainable management?* New York: Columbia University Press, xiii–xvii.

World Commission on Environment and Development (WCED), 1987. *Our common future.* Oxford: Oxford University Press.

Zarin, D.J., *et al.* eds., 2004. *Working forests in the neotropics: conservation through sustainable management?* New York: Columbia University Press.

Amazonian agriculturalists bound by subsistence hunting

Eric Minzenberg[a] and Richard Wallace[b]

[a]Earth Sciences Department, Santa Monica College, Santa Monica, CA, USA;
[b]Department of Anthropology & Geography, California State University Stanislaus,
Turlock, CA, USA

As rural Amazonian *caboclo* households transition from subsistence to
increasingly market-based economies with systems of individualized
production and consumption, the non-market exchanges of cooperative labor and game meat integral to the practice of subsistence hunting
maintain bonds between households in the community. Further, subsistence hunting strengthens kinship ties across households wherein both
men and women are important actors, an additional important component of local resource management. This study conducted in Brazil's first
sustainable development settlement, *Projeto de Desenvolvimento Sustentável (PDS) São Salvador*, suggests that not only are use rules important
in the local management of natural resources, but also that inter-
household relations of social conflict and social cohesion expressed in
subsistence hunting reinforce community social structure while simultaneously impacting the natural resource base. In the effort to link social
with ecological sustainability, conservation and development planners
must recognize that extra-community government regulation may inter-
fere with the socio-cultural dynamics of local community identity and
governance explicit in subsistence hunting in rural Amazonian *coboclo*
communities and could potentially produce negative consequences for
community social structure and the natural environment.

Introduction

Amazonian *caboclos* have historically been portrayed as extremely poor,
invisible, tucked away in the Amazon hinterlands, and disengaged from
modernity (Adams *et al.* 2009; Nugent 2009). *Caboclos,* river dwelling
populations, are descendants from the miscegenation of indigenous
peoples with the Portuguese, and to a lesser extent, the miscegenation
of indigenous Amazonians with the descendents of African slaves from
Northeast Brazil who migrated to work in the debt-peonage, patron-client
Amazonian rubber economy (*aviamento*) at the turn of the twentieth

century (Wagley 1953; Ross 1978; Parker 1985; Schmink and Wood 1992; Nugent 1993). Challenging this portrait of invisibility, recent scholarship has recast the identity of *caboclos* as both active participants in production and markets, adaptive and flexible in managing their connectivity with modernity actively creating their futures and forging their identity (Brondizio 2009; Harris 2009; Nugent 2009) while injecting new values into forest landscapes (Brondizio 2009). *Caboclos*, and rural Amazonian populations more generally, both through individual household initiatives, government policy, and the arrival of regional conservation and development projects, are increasingly integrated into the market economy. This transition involves a shift from non-market community exchanges in subsistence (e.g. hunting, labor exchange), and traditional market-oriented extractive activities (e.g. rubber extraction) with patron-client ties, to an economic and social system predicated on market-oriented agricultural production, now managed by individual households (Salisbury and Schmink 2007). As a result, there is increased concern about how local households negotiate this transition and its implications for the Amazon rainforest (Godoy *et al.* 1997).

The *caboclo* economy combines subsistence production with marketing of surplus production in local and regional market centers (Parker 1985; Redford and Padoch 1992; Harris 2000) and hunting remains integral to the forest-based activities of Amazonian *caboclo* populations. Rubber tapping, small-scale agriculture, animal husbandry and other sustainable activities of rural peoples would not be possible without subsistence hunting (Redford 1992, p. 421). Game meat frequently supplies a large portion of the protein and caloric intake for rural households in the developing world, while also holding elevated status in the culinary tastes of rural peoples (Harris and Ross 1987). That is, game meat is both a delicacy and has greater social significance for rural peoples than other foods. Hunting not only provides protein and calories for rural Amazonians (Robinson and Bennett 2000; Bennett 2002; Rao and McGowan 2002), but also takes on social functions as well, in the cooperation or competition over wild game (Aspelin 1979; Kensinger 1983; Kaplan and Hill 1985; Stearman 1989; Hawkes 1993; Kelly 1995; Minnegal 1997; Gurven *et al.* 2000; Patton 2005; Hames and McCabe 2007).

Harris (2000) writes that in *caboclo* communities in Amazonia, kin desire to reside in clusters of households within close proximity to one another, and reciprocal distribution of fish and wild game between households creates a sense of community and affirms the value of these relationships. Inter-household conflicts are often expressed as quarrels between kin groups. Links between households are often maintained by women in the community who distribute the products of fish and game derived from the fishing and hunting efforts of men.

In Amazonia individuals from multiple households partake in the non-market exchange of game meat between rural households (Aspelin

1979; Kaplan and Hill 1985; Hawkes 1993; Gurven *et al.* 2000; Hawkes *et al.* 2001; Hames and McCabe 2007). Conversely, market transactions that entail the transfer of money for goods and/or services generally involve only the buyer and seller, a mechanism that does not reinforce inter-household bonds. Thus, as rural households transition to increasingly market-based economies with systems of individualized production and consumption, non-market, inter-household exchanges of labor and meat integral to the practice of subsistence hunting help extend and maintain the bond between households in the community (Minzenberg 2005).

Within the local community, social conflict and social cohesion are often expressed through hunting practice. Although social conflict often frays inter-household relationships, it is also an important mechanism of communication across households (Simmel 1955; Coser 1956). Social conflict refers to the conflict between an aggregate of individuals as opposed to conflict between single individuals (Oberschall 1978). It results from the actual or perceived competition over resources, values, interests or prestige (Oberschall 1978). Social cohesion via inter-household exchanges brings households together and acts to contain conflict. Social cohesion can be viewed as both a process of managing conflict between aggregates of individuals, and the institutionalized practices that contain potential conflict. These processes and acts display group participation and commitment (Gudeman 2001, p. 36), thus shaping group identity.

Political ecologists have also made important contributions to understanding how households and communities are influenced by economic and political forces (Blaikie 1985; Schmink and Wood 1987; Peluso 1992; Peet and Watts 1993; Stonich 1995; Bryant and Bailey 1997; Watts and Peet 2004). This includes scaled analysis to examine conflicts over rights and access to resources (Yeh 2000; Nygren 2004; Robbins 2004; Roth 2008), including wild game (Povinelli 1992; MacDonald 2005; Blaser 2009). Hecht (2004) argues that it is critical to consider how local communities negotiate these forces and mediate change. She suggests the need to explore various explanations to understand forest persistence and how communities adapt to environmental change and conflict at the micro level, including the investigation of how local livelihoods and institutions shape political-economic transformation. Neumann (1992, p. 87) argues for an approach that builds from the bottom-up, examining local actors and social relations of production, how these actors are linked to "wider geographical and social settings" and nests the present within a historical context. This approach allows for a more critical understanding of how rapid economic and social transformations create tension between short-term, impersonal market exchanges and traditional community exchanges built through long-term, social relationships (Gudeman 2001).

In this article, we examine the livelihood transition since the early 1990s for the *caboclo* residents of Brazil's first sustainable development settlement, *Projeto de Desenvolvimento Sustentável* (PDS–Sustainable

Development Project) *São Salvador*, in the western Amazonian state of Acre. We explore the social importance of hunting for settlement residents whose livelihoods have transitioned from subsistence and rubber tapping to market-based agricultural production. We argue that subsistence hunting in the settlement binds households together through conflict and cohesion within an increasingly individualized market economy. The practice of subsistence hunting involving non-market exchanges between different households acts to reconstitute community and reinforce group identity and serves to reinforce familial ties across households wherein both men and women are important actors.

Historical development of *PDS São Salvador*
Rubber economy—1930s to 1990

The dynamics of the debt-peonage *aviamento* rubber system (Dean 1987; Barham and Coomes 1994; Coomes and Barham 1994) were much the same in *Seringal São Salvador*[1] as elsewhere in Amazonia. However, rubber production began much later in the area, well after the initial peak of rubber activity throughout Acre at the turn of the past century, commencing only in the early 1930s. The individuals that were recruited to tap rubber in the *seringal* were mostly born in western Acre, although many of their parents were part of the wave of immigrants that came to Acre from the northeastern Brazilian state of Ceará. Each *seringueiro* (rubber tapper) was required to pay an annual land-use fee of 50 kilograms of rubber to the *patrão* (Pedro de Morais). Da Cunha and de Almeida (2000) estimate that in the Juruá Valley in western Acre in the 1970s, each *seringueiro* household produced, on average, 600 kilograms of rubber annually giving 10% as payment to their *patrão*.

In the mid 1980s, the Morais family abandoned *Seringal São Salvador* as the profitability of rubber production throughout Brazil continued its steady decline. The price of rubber in the nearby Alto Juruá in western Acre fell from US $1.80/kg in 1982 to US $0.40/kg in 1991 (Pantoja 2004). By the late 1980s, the residents of *Seringal São Salvador* were left to their own devices, pursuing whichever livelihood strategies best suited their households and families. Usually this involved a mixture of subsistence fishing and hunting activities combined with occasional marketing of rubber and agricultural products.

An autonomous rubber system, wherein each household controlled their own production and sale of rubber without the control of a *patrão*, briefly developed in the *seringal*, but was quickly eclipsed when the subsidies for rubber were relaxed by the Brazilian government in the 1980s, and effectively abandoned in the early 1990s. Rubber lost its central importance in the socio-cultural and economic livelihoods of the residents of the area, and by 2005 none of the residents of *PDS São Salvador* tapped rubber as a means of procuring a livelihood. A minuscule minority of residents continued to tap rubber to supplement their income from other economic pursuits. The shift in household livelihoods from rubber to small-scale agricultural production created difficulty for some of the older residents of *PDS São Salvador* in their engagement with a new economic system that had broken from the traditional patron-client relationship established by the *aviamento* rubber system.

Transition to agriculture (farinha)—1990 to present

In the post-rubber rural economy in *Seringal São Salvador* in the early 1990s, residents went through a quick succession of household livelihood strategies initially relying principally on timber harvesting, then moving to the sale of wild game and animal skins[2] and finally to reliance on small-scale, family based agriculture centered on the production of *farinha* (flour made from the cassava root) for market (Câmara 1999). *Farinha,* historically the staple of Amazonian diets, became the main source of cash income for the majority of the residents of *Seringal São Salvador.* Accompanied with this shift in livelihood was a corresponding transformation of resident identity from rubber tapper to agriculturalist.

Seringal São Salvador was acquired by the Brazilian Colonization Agency, *Instituto Nacional de Colonização e Reforma Agrária* (*INCRA*), in 1999 to be used as a resettlement project for families living within *Parque Nacional Serra do Divisor* (*PNSD*) (PESACRE 1998). In 1989, PNSD was created on the western border of *Seringal São Salvador,* abutting the frontier with Peru (Figure 1). The abandoned *seringal* contained nearly 28,000 hectares of forest bordering *PNSD,* and consequently this area is

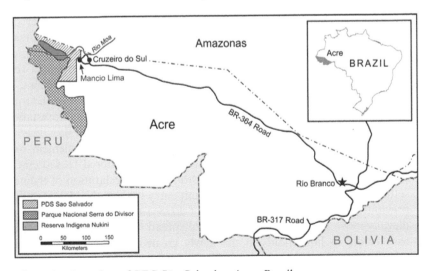

Figure 1.   Location of PDS São Salvador, Acre, Brazil.

an important link into the overall scheme of buffer zone management of the national park. The use of wildlife by the residents of *PDS São Salvador* and the judicious management of faunal resources by inhabitants in this area was of critical importance in the conservation and development agenda, regional planning process, and in the creation of the new settlement (Câmara 1999). *INCRA's* original plan was to resettle families living within the border of the *PNSD* into family-owned small farms within *Seringal São Salvador* (PESACRE 1998). *INCRA* concluded that the resettlement project was not advisable, and could potentially have negative environmental and social impacts on both the land and the families living there.

Anxious residents of *Seringal São Salvador*, fearing that the government would move more families into the seringal, were brought into dialogue with *INCRA* with the help of a Brazilian NGO, *Grupo de Pesquisa e Extensão em Sistemas Agroflorestais do Acre* (*PESACRE*), to discuss resource management options within the 28,000 hectares of tropical forests. In the process of negotiation that resulted from the collaboration of Brazilian federal agencies, PESACRE and local residents, *Seringal São Salvador* became *Projeto de Desenvolvimento Sustentável* (*PDS, Sustainable Development Project*) *São Salvador* on 14 July 2000. Thus, *PDS São Salvador* became the first sustainable development settlement created in Brazil.

Within this new settlement, the federal government retained title to the land and resident households were given 20-year renewable usufruct leases. Each household was given title to usage of a 100-hectare lot (with joint ownership shared by husband and wife), although by law, only 20% could be cleared for agricultural use and livestock production. Although households were grouped into communities, agricultural lots were managed by individual households with all production from this land under the ownership of the lot designee. Animal, plant, and botanic resources outside the bounds of each 100-hectare lot within the settlement were common property, open for harvest by all residents of the settlement. Provisions were made for future generations (future lot grantees) within the settlement, as long as deforestation did not reach more than twenty percent of the total area of the settlement. It was estimated in 1999 that current deforestation within the settlement was less than five percent (Câmara 1999).

One hundred seventeen families in the newly-named settlement registered with *INCRA* in 2000 to receive the credit offered to settlement residents by the federal government. Although the declaration of the new settlement on 14 July 2000 did not permit entry beyond this date, families previously living outside the bounds of *PDS São Salvador* continued to illegally move into the area. When word spread that *INCRA* was offering easy credit to households in the newly created settlement, some poor families from the closest town of Mâncio Lima scrambled to set up shelter in the *seringal* so as to acquire money and a piece of land to live and work.

By January 2003, *INCRA* listed 150 agricultural lots, each registered with a different family, ranging in size from a low of 17 hectares to a high of 35 hectares.

In the process of constructing the sustainable development settlement, *INCRA* required households to assemble themselves within named communities. Historically, households in western Acre were grouped within small family clusters, as was common for many rural *caboclo* households in the Brazilian Amazon (Harris 2000), from as few as two households to ten households. In the resulting process, ten communities were established, most taking the old names of the *colocações* (forest holdings that each household tapped to produce rubber) during the rubber era. Within each community, this new settlement pattern joined generally three to seven different family groups together into one loosely aligned, sociopolitical grouping. The process of shifting from a familial-based social-political structure to a community-based one was a difficult transition for most households in the settlement, with community disputes and resolution frequently defined on the basis of familial association.

In the two years immediately following the declaration of *PDS São Salvador* in 2000, PESACRE, in conjunction with the residents of the ten communities of the settlement, defined a series of regulations and management goals in regards to the use of natural resources within the settlement (Comunidade do Projeto de Desenvolvimento Sustentável São Salvador 2003). Community regulations set guidelines for residents in regards to hunting, fishing, timber exploration, animal husbandry, and the clearing of forestland. The aspect of resource management that drew the most attention was hunting, with nineteen regulations created to regulate hunting practice within the settlement (See Table 1). A common trigger point of animosity between households within communities in *PDS São Salvador* remained hunting practice, and the use of wildlife.

Methodology

The analysis here is based on fieldwork conducted by Minzenberg in *PDS São Salvador* from July 2003 through April 2004. Household surveys collected basic socio-economic data for all household members with all adult household residents (individuals ≥ 15 years old) present for the group interview. This occurred in 59 of the 64 households that existed in six of the ten communities in the settlement. Adult householders were asked if they hunted or not, if they traded labor agricultural labor days with their neighbors (*troca dias*), if they engaged in hunting outings with members of other households, if they shared wild game with other households in the community and if so, with which households, and which gender in the household was responsible for the distribution of wild game to other households in the community. Householders were also asked to

Table 1. Community hunting regulations in *PDS São Salvador*.

Hunting *porquinho* [also commonly called cutia, *Tayassu tajacu*] with dogs is prohibited.
Hunting with dogs is prohibited.
Dogs are permitted to accompany residents to keep wild animals from attacking residents' domesticated animals.
Hunting for food for a trip is permitted. Only 1/2kg per person per day up to a maximum of 20kg.
Hunting the offspring of deer is prohibited.
If you see a track of offspring of deer you are not permitted to shoot at the offspring.
If you see tracks of the offspring of deer or armadillo, do not kill the mother.
Hunting to feed oneself is permitted.
Hunting animals threatened with extinction is prohibited.
Minors can only hunt with permission from their parents.
You must advise your neighbors when setting a trap [*armadilha*].
Traps are permitted only from 5pm until 7am the next morning.
Traps must be accompanied with a warning sign advising of their use in the vicinity.
A fence must be put up around an *armadilha*.
Traps are to be set up in the buffer of one's community.
Hunters that do not live in the settlement are not permitted to hunt.
Hunting to raise the offspring of wild animals is not permitted.
Hunting to sell the meat of wild game is not permitted.
Hunting turtles is permitted.

*Translated by Minzenberg (from Portuguese) from Comunidade do Projeto de Desenvolvimento Sustentável São Salvador (2003).

describe conflicts in the community that resulted from quarrels over hunting practice since the establishment of the settlement in 2000.

An in-depth hunting outing survey measured the hunting practices of four households in one community and three households in another over a period of four months (January 2004 through April 2004). The households selected to participate in this exercise were based on the results obtained from the household questionnaire explained above. Specifically, these seven households contained individuals who hunted throughout both the rainy (December through May) and dry (June through November) seasons. Furthermore, the adult members of these households lived in *PDS São Salvador* throughout the rainy season months, allowing for monthly sampling of hunting outings by members of these seven households. In contrast, adult members of other households in the community often migrated to live with family in urban areas for some time during each of the rainy season months.

Hunting households were tracked for seven consecutive, randomly selected days per month for four consecutive months. At the end of each of the seven consecutive days hunters were interviewed about their hunting outings. Hunters that made extended-stay, overnight hunting trips were interviewed when they returned to their households and communities, following the conclusion of their hunting activity. Hunters were asked if they hunted alone or in groups, which individuals were part of the hunting group, the duration of the hunting outing, when and where they hunted, animals killed during the hunt and by whom in the hunting party.

Ethnographic data, including observations of meat exchange between residents of different households at the conclusion of successful hunting outings, supplemented the information collected from the surveys described above. In addition, Minzenberg attended each of the monthly meetings of the settlement's resident advisory council (*Conselho Gestor*) from September 2003 to April 2004. The council, comprised of resident leaders of each of the ten settlement communities organized by PESACRE, discussed resource management issues and defused community disputes.

Results

Forty-nine of the 59 households surveyed in six communities in the settlement were hunting households (e.g., at least one adult member of the household hunted). The average number of times residents of these households hunted per month was 2.08 (sd = 1.85), ranging from zero times per month to a high of ten times per month in one household. Many hunters stated they hunted more in the past, but currently their agricultural responsibilities did not afford them adequate time to hunt.

Gender played a key role in hunting practices within the settlement. As was common in the days of the rubber economy in Amazonia, both men and women continued to perceive the forest environment (where hunting occurred) as a masculine space (Wolff 1999; Pantoja 2004). One woman stated, "It is ugly for women to walk in the forest,"[3] and a man from another community said, "Men work in the forest. Women in the house." Men gave varied reasons for hunting including a wish to escape the "*barulho*" (noise) of the household, that "the forest helps people" because it is "calm in the forest," that hunting is part of "the culture of the region," and that he hunts because "it is the *jeito* I have." The word *jeito* as used by Brazilians has a variety of meanings in English that may include some of the following, context dependent: "way," "manner," "knack," "skill," "appearance," "personality," and/or "style." Hunting, therefore, was decidedly a male activity as only one woman in these 49 hunting households hunted, yet only a few times per year as opposed to a number of times per month for men who hunted.

Conflict-hunting and wildlife management in PDS São Salvador

In the years that followed the establishment of norms of resource use in the settlement, hunting disputes were the catalyst of community conflict

107

(Vângela Nascimento, biologist, PESACRE, personal communication, 10 May 2004).

Lingering disputes within individual communities over hunting territory and practice often pitted one family group against another. In one community in the settlement, one familial group accused another of hunting with dogs. Hunting with dogs was historically a common practice in the *seringais* of Acre (Melo 2000). Within the *PDS São Salvador* settlement, however, the natural resource use plan did not permit hunting with dogs. A hunter from another community blamed the paucity of animals in the settlement on the use of hunting dogs. He said that after the fall of the rubber economy in the area by the early 1990s, households turned to the sale of wild game as the principal means of household income, with hunters principally using their dogs in the *seringal*. As a result of the success in hunting prey with dogs, there were fewer animals in the forest in 2005. Fragoso *et al.* (2002) found that some wildlife species had become locally extinct as a result of hunting pressure by residents of the area.

Other hunting methods sparked conflict between households in the settlement. Disagreements over the placement of *armadilhas* (traps consisting of the sawed-off barrel of a rifle with a trigger that can be tripped by small animals) within the community frequently served to inflame tensions. *Armadilhas* were used to hunt small animals that burrowed into downed, rotted logs, or directly into a hole in the forest floor, commonly paca (*Agouti paca*) and different species of armadillo (*Dasypus sp.*).

The bounty of the hunt, wild game, frequently served to intensify animosity between competing family groups when meat was not ex-changed across these boundaries. Traditionally, expectations within rural communities throughout Acre required the sharing of game meat (Calouro 1995; Melo 2000; da Cunha and de Almeida 2002; Minzenberg 2005). Some individuals in *PDS São Salvador* complained that other households in their communities did not share wild game with them. In one community, two different family groups, though related through inter-marriage, were feuding and did not reciprocate in the partitioning of wild game with each other. One woman in this community complained that the other family group sometimes hid their successful hunt from her family group so as to not be burdened with sharing meat with their neighbors. In another community, a sister from one household was arguing with her brother from another household for several days. One evening the brother returned from a hunt with a few birds and a small monkey. His wife gave meat to the households of four of his five siblings who lived in the community, to the exclusion of the household of the sister with whom he was feuding. His spouse, in support of her husband, said of her sister-in-law, "She is not going to eat today." A common manner of showing displeasure and of punishing members of another household was to refuse to share hunted game with the offending household members.

It has been reported elsewhere in Amazonia that men from different households would sometimes give game meat to their female lovers living in other households within the community (Chagnon 1968; Siskind 1973; Kensinger 1983). Siskind (1973) called this the reciprocal exchange of meat for sex. In one community in *PDS São Salvador*, residents complained that this exchange occurred between a married man of one household and his married female lover from another household. Family of the spouses of the adulterous couple stated that the adulterous man would only hunt when his female lover requested that he bring her meat. They said that he was generally lazy, but quickly responded to the demands of his female lover. The adulterous couple was eventually forced out of the community by the families of the jilted wife of one household and jilted husband of the other.

An additional problem emerged in selling wild game in nearby city centers, a violation of Brazilian federal law. A resident of the settlement told Minzenberg that along the Moa River (which bisects *PDS São Salvador*) during the rubber era, everyone hunted to sell wild game. Although illegal, at the end of 2003 wild game was selling in the city of Mâncio Lima for around $R5 per kilogram. In 2004, the two areas of the settlement rife with tension resulting from accusations of illegal game sales were in the northwest border with the Nukini Indigenous Reserve, and in the area of the northeast border with private landowners. Frequent encroachment by individuals hunting to sell game, who were not inhabitants of the settlement, occurred in these boundary areas. Many species of wild animals were considered delicacies in the region including paca (*Agouti paca*), white-lipped peccary (*Tayassu pecari*), and red brocket deer (*Mazama americana*), and there were willing buyers of the meat from these animals in the nearby city centers. Several residents of the settlement stated they had been threatened with "a bullet in the back of their head" if they denounced these illegal hunters to the federal environmental protection agency, *Instituto Brasileiro do Meio Ambiente e dos Recursos Naturais Renováveis (IBAMA)*. According to officials in the IBAMA office in the city of Cruizeiro do Sul, wild game sales occurred mostly from hunters who lived outside the settlement, less frequently by residents of *PDS São Salvador*.

Cohesion–social exchange through hunting

As residents of *PDS São Salvador* had little experience with self governance, the *Conselho Gestor* proved to be an important arena for bringing community leaders together to discuss the management of settlement resources.[4] Overall, two issues dominated discussion at these meetings: concerns over hunting practices and a long-standing feud in one community between two households (different families) over agricultural plots which finally ended when one of the families peacefully conceded to the demands of the other family. Every dispute aired over the use of natural resources featured settlement residents against non-settlement residents or one family group against another in a single community in the settlement. Although each of the meetings was contentious, these meetings offered an opportunity for residents to debate the present and future management of natural resources in the settlement and sometimes reach agreement on a course of action. On two occasions Minzenberg accompanied a member of the *Conselho Gestor* to the offices of IBAMA in the city of Cruzeiro do Sul with a written letter from the *Conselho* denouncing illegal resource use in the settlement. Residents of the area did not dare denounce illegal resource use during the era of the rubber economy for fear of financial and bodily harm at the hand of the Morais family.

During the late winter months of January through April, a period during the year described locally as *mau de rancho*, or "bad for food," food resources were harder to procure for settlement residents. Agricultural production was low during the wet season and rising water levels flooded the forest giving fish ample room to hide, decreasing the catch of fish from the rivers in the settlement. Consequently, male hunters sometimes resorted to overnight, extended-stay hunting trips lasting upwards of one week as a means to secure enough food to last several days for their households. Our study showed that extended-stay hunting outings were more frequently made with kin residing in other households than with non-kin hunting partners. More specifically, four of five hunting outings involved other kin. Likewise, single-day group hunting parties during this four-month period in these two communities normally consisted of kin groups (11 of 17 hunting outings, or 64.7%). Figure 2 shows the kin relationships of the individuals in these hunting groups.

Extended-stay hunting trips afford the opportunity for binding friendships among men of different households within the community. Often men who traded days of labor in agriculture with one another (*troca dias*) also made extended-stay hunting trips together. In *PDS São Salvador*, households that trade days of labor with one another were usually family-related. A Pearson correlation showed a statistically significant positive relationship between hunting frequency (measured as the number of hunting outings per month) and trading labor days with other households in the community ($n = 59$, $r = .309$, $p \leq .017$).

Meat exchange

Game meat, traditionally the most appreciated food in the *seringal* (Wolff 1999), held elevated status in the food preferences of residents of the settlement. An adult resident from one community remarked that "meat is

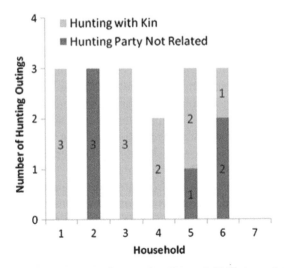

Figure 2.   Single-day group hunting parties (January 2004 through April 2004).

the best food." Frequently when Minzenberg inquired what a particular household was having for dinner, a common response if they did not have any meat (or fish) to eat was "nothing, just *farinha*." *Farinha* was the staple of the diet in *PDS São Salvador*, and no meal was complete without it, but wild game held special significance within the dietary preferences of the household. Animal meat was served during *festas* (parties) for the inaugurations of schools, churches, or during other congregations of households and communities in the area. No one would consider serving just fish and *farinha* to their guests. Hunters in hosting household(s) would often hunt in the days preceding the *festa* in order to have game meat to serve to their guests.

As Marcel Mauss (1990) wrote of the exchange of gifts, the exchange of meat between settlement households involved a three-part process—to give, to receive, and to reciprocate. This social system of exchange operated with households acting in all roles—giving, receiving, and reciprocating. Meat exchange helped to insure the survival of all members of the community as this rotating system of consumption helped feed different members in the community, especially important during the *mau de rancho* winter months.

Households in the settlement partitioned game meat to other households based on three factors: (1) the individuals involved in the hunting party, (2) the size of the kill, and (3) kin relationship between households. Any successful kill from a hunting outing was first equally divided amongst the hunters in a hunting party. If enough meat remained to give to other households in the community, meat was further subdivided to selected households nearly always related by blood or marriage.

Most hunting households stipulated that the amount of meat they gave to other households depended on the size of the animal. If a hunter returned to the community with a small animal (called *embiara*) such as black agouti (*Dasyprocta fuliginosa*) or paca (*Agouti paca*), or one of the commonly hunted birds in the seringal, including Spix's Guan (*Penelope jacquaou*), pale-winged trumpter (*Psophia leucoptera*), or white-throated Tinamou (*Tinamus guttatus*), his family kept the entire animal. These hunts provided a single meal for the family. Often the hunter returned to the forest to hunt the following day, to put food on the table. With larger animals including white-lipped peccary (*Tayassu pecari*) or red brocket deer (*Mazama americana*), the household of the successful hunter would either keep one-half (*uma banda*) or one-fourth (*um quarto*) of the animal. Whether the household kept one-half or one-fourth of the meat of the hunted animal depended upon how many households in the community were connected to the meat-giving household within the system of reciprocal exchange of wild game. Kin were usually the channel of meat exchange and thus, an important part of the channel of social relations within the community. In the sample of 49 hunting households in the settlement, 80% of hunting households exchanged meat only with kin-related households. (See Figure 3.)

Women in peasant Amazonian communities also play important roles in this system of inter-household relationships (Harris 2000), further reinforcing familial and community connections. Typically, men hunt wild game and bring it back to their households, and women cut, clean, cook, and partition meat to their own, and frequently other, households in the community. Of the 49 hunting households in this sample, women participated in distributing wild game to other households in 35 (71.4%) of the hunting households. Figure 4 shows the gender breakdown of meat division within hunting households.

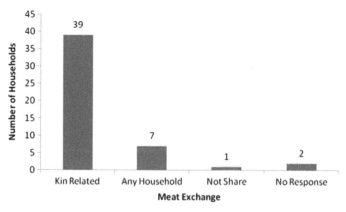

Figure 3.   Meat sharing by relation to giving household in 49 hunting households.

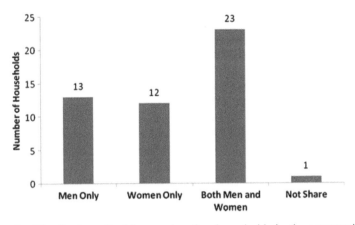

Figure 4. Distribution of wild game to other households in the community by gender (in the giving household) within 49 hunting households.

Discussion

*PDS São S*alvador was created in the search for alternatives to stem the tide of destructive land-use practices that have plagued much of Amazonia in the past. In doing so, a new strategy joining governmental agencies with Brazilian NGOs and local residents of the settlement sought to account for the ecological diversity of the region while also respecting the cultural and social histories and livelihoods of inhabitants in the area. Rural peoples have been key players in the agenda of the Forest Government of Acre, Brazil (see Vadjunec *et al.* 2011) that seeks to encourage rural Acrianos to continue to create sustainable and prosperous livelihoods, while refraining from migration to urban areas (Kainer *et al.* 2003). Wild game is an important food source for rural dwellers in Acre (Calouro 1995; Zoneamento Ecológico-Econômico do Acre 2000) and resource managers have worked to maintain healthy and viable populations of wild animals throughout the state. Community meetings conducted over several years established rules for the use of water, botanical, and animal resources in the settlement. Today the challenge facing *PDS São Salvador* is the implementation of these regulations coupled with the continuing search for ecological and social sustainability for future generations. This is a daunting task given that the inhabitants of the settlement are among the poorest in the poor state of Acre (Shaeff 2002), relying in large measure on nature's bounty for their sustenance, and are a young population with population pressure only to increase in the years to come (Câmara 1999).

Hunting practice has become important in *PDS São Salvador* in the maintenance of a group identity within communities that have witnessed a dramatic and rapid livelihood shift from rubber tapping to small-scale agriculture. With the decline of the patronage system by the late 1980s linking *patrão* (Morais family) and *seringueiro* (rubber tapper), kinship ties were the glue that bound households together within the community. As households rapidly transitioned to systems of individualized market agricultural production, hunting practice has become a principal mechanism through which kinship is enacted and households are tied together. These linkages act as safety nets in times of need (Scott 1976; Ellis 1993; Ellis 2000). For example, meat sharing during the *mau de rancho* winter months in *PDS São Salvador* when food resources are difficult to acquire, and these intra-household connections act as a regulatory mechanism in the exploitation of game resources in the settlement (e.g., Hecht 2004).

It is through conflict and cohesion associated with hunting practices and the use of wildlife in *PDS São Salvador* that relations across *caboclo* households are strengthened. Social conflict creates tension which can tear apart group solidarity, yet it can also be constructive in that householders, through this form of community engagement, reconstitute group formation and cohesion (Simmel 1955; Coser 1956). This was clearly seen in the monthly settlement meetings of the *Conselho Gestor* that were important forms for dispute resolution over the management of wild game and other natural resources. Reciprocal relations between households are performed through the practice of non-market hunting exchanges (social cohesion) and alternatively, signified via the absence of non-market exchanges between households in the community (social conflict).

Conservation strategies that build upon community-wide norms of resource use may forestall more individualized land-use practices such as cattle ranching that are socially and environmentally destructive. Already in one community in *PDS São Salvador* where residents have frequently refused to participate in meetings with PESACRE to strengthen adherence to settlement resource rules (see Table 1 above), householders have increasingly turned to cattle production as a livelihood choice (Salisbury 2002). Wallace (2004) found in his study of rubber tappers in the Chico Mendes Extractive Reserve in eastern Acre that households invested a greater portion of their wealth in cattle as they became wealthier. Cattle rearing does not bode well for the conservation of Amazonian environments, as increasing cattle production has been linked with increases in deforestation (Hecht and Cockburn 1990; Loker 1993; Fearnside 1997).

Rural livelihoods in *PDS São Salvador* reflect a mixture of market and non-market transactions and as such, are fundamentally different from capitalist livelihoods that are predicated on the market. Hunting plays an important role in the creation of these non-market exchanges. Meat sharing, hunting with other households, and trading labor days are all non-market transactions between households that are an essential part of the hunting strategy in *PDS São Salvador* (Minzenberg 2005). Whereas the marketing of agricultural products often involves individual household livelihood strategies, hunting practice in the settlement entails group livelihood strategies joining several households in the community.

Community social structure is reinforced by these non-market mechanisms enacted in hunting practice that bind households together more so than via individualized market transactions.

Conclusion

This study carries important implications for conservation and development of the Amazon region, and more specifically conservation areas, such as extractive reserves and other sustainable development areas. It suggests that strategies imposed from outside the local community that heavily control hunting practice as a mechanism to conserve regional wildlife could serve to weaken kinship ties in the community that are critical to reciprocal exchanges across households. Group identity can play an important role in conservation and development strategies in the region. Elsewhere, kinship ties have been shown to reinforce community cooperation in the management of natural resources (Begossi 2001; Lu 2001). Meat substitutes for wild game such as animal rearing by individual households may be ecologically attractive, but may weaken inter-household ties at the local level. The raising of cattle in Amazonia has historically proved disastrous for local communities as well as the natural environment (Davis 1977; Anderson 1990; Hecht and Cockburn 1990; Almeida 1992; Diegues 1992; Schmink and Wood 1992; Schwartzman 1992). Inter-household networks can be critical to the survival of individual households and the community as a whole (e.g. game meat sharing during *mau de rancho* months), and chipping away at the bonds created in subsistence hunting may change important socio-economic relationships in rural communities.

Thus, our research supports arguments noted by others that management of common property resources such as wildlife will only be successful if local peoples have an active role in the administration of these resources (Ostrom 1990; Gibson *et al.* 2000; Burger *et al.* 2001; Ostrom *et al.* 2002). Tropical forest animals are key components influencing the structure and function of tropical forests (Redford 1992; Robinson and Bennett 2000; Meijaard *et al.* 2005), thus hunting practices and wildlife management by local communities is of critical importance to the ecology of the tropical forest environment. Conservation and development planners must include local peoples in the management of wildlife as "hunting issues cannot be separated from land use" (Wadley *et al.* 1997, p. 263).

Further, this study argues for the need for ethnographic research to ensure a thorough understanding of the emergence of socio-cultural institutions in local contexts and the critical role they plan in community resource management (Hecht 2004). In *PDS São Salvador*, ethnographic household research revealed how social relationships are integral in subsistence hunting and wild game use in the local livelihood system.

In-depth household study provided an important local window for understanding the social dynamics involved in the practice of hunting within a transitioning economy, how both conflict and cohesion emerge through this practice, helping forge both community identity and resource use.

Amazonian landscapes, like their inhabitants, are not static, but dynamic, both culturally and ecologically, historical products of both the material needs and spiritual values of local peoples. As new programs such as REDD (Reducing Deforestation and Degradation) (Nepstad *et al.* 2007; Angelsen and Wertz-Kanounnikoff 2008), and the Forests Assets Program of the State of Acre (State of Acre 2009) focus on reducing deforestation and valuing local livelihood systems for their role in reducing carbon emissions, a more complex and local understanding of the livelihood practices and institutions that help keep forests standing is critical. Policy makers and NGOs involved in implementation and stewardship of these programs should be cognizant of the complex social relationships and institutions that are embedded in and govern the use of local landscapes. This understanding is critical when evaluating the potential success and impact of conservation and development initiatives that are implemented in local cultural and institutional contexts. It is important for valuing both the cultural and biological diversity of Amazonia.

Acknowledgements

Funding for this research was provided by the National Science Foundation (Doctoral Dissertation Improvement Grant, #6391954) and Working Forests in the Tropics Small Grants Competition at the University of Florida. The Brazilian NGO, PESACRE, was very accommodating in providing logistical support during Minzenberg's stay in Acre. We especially thank biologist Vângela Nascimento and agronomist (Cazuza) Eduardo Borges of PESACRE. Thanks to Michael Larson for his help in preparing the map of *PDS São Salvador*. We would also like to thank three anonymous reviewers and the following individuals for their helpful comments on earlier drafts of this paper: Laura Bird, John Cinnamon, Cameron Hay-Rollins, Jonathan Larson, Michael Minzenberg, Marianne Schmink, Jackie Vadjunec, and Tracy Van Holt.

Notes

1. Prior to being named the first sustainable development settlement within Brazil in 2000, *Projeto de Desenvolvimento Sustentável (PDS) São Salvador* was commonly referred to as *Seringal São Salvador* in deference to its use and designation as a rubber tapping estate. The word *"seringal"* is used in Brazil to designate a current, or former, rubber estate.
2. The sale of wild game and other animal products has been prohibited in Brazil since the declaration of federal law 5.197 in 1967.

3. All quotations from residents of PDS São Salvador were translated from Portuguese by Minzenberg.
4. Interestingly, the two communities in the settlement with householders who owned the largest herds of cattle were the only communities that did not once send representatives to meetings of the *Conselho Gestor* during the meetings that Minzenberg attended from September 2003 to April 2004.

References

Adams, C., Murrieta, R., and Neves, W., 2009. Introduction. *In*: C. Adams, *et al.*, eds. *Amazon peasant societies in a changing environment: political ecology, invisibility and modernity in the rainforest*. Doetinchem, Netherlands: Springer Science and Business Media Netherlands, 1–18.

Almeida, A., 1992. *The colonization of the Amazon. Austin*. Texas: University of Texas Press.

Anderson, A., 1990. *Alternatives to deforestation: steps toward sustainable use of the Amazon rain forest*. New York: Columbia University Press.

Angelsen, A. and Wertz-Kanounnikoff, S., 2008. What are the key design issues for REDD and the criteria for assessing options. *In*: A. Angelsen, ed. *Moving ahead with REDD: issues, options and implications*. Bogor, Indonesia: CIFOR, 11–21.

Aspelin, P., 1979. Food distribution and social bonding among the Mamaindê of Mato Grosso, Brazil. *Journal of Anthropological Research*, 35 (3), 309–327.

Barham, B. and Coomes, O., 1994. Wild rubber: industrial organisation and the microeconomics of extraction during the Amazon rubber boom (1860–1920). *Journal of Latin American Studies*, 26 (1), 37–72.

Begossi, A., 2001. Cooperative and territorial resources: Brazilian artisanal fisheries. *In*: J. Burger, *et al.*, eds. *Protecting the commons: A framework for resource management in the Americas*. Washington D.C.: Island Press, 109–131.

Bennett, E., 2002. Is there a link between wild meat and food security? *Conservation Biology*, 16 (3), 590–592.

Blaikie, P., 1985. *Political ecology of soil erosion in developing countries*. New York: Longman Group Limited.

Blaser, M., 2009. The threat of the Yrmo: the political ontology of a sustainable hunting program. *American Anthropologist*, 111 (1), 10–20.

Brondizio, E., 2009. Agriculture intensification, economic identity, and shared invisibility in Amazonian peasantry: caboclos and colonists in comparative perspective. *In*: C. Adams, *et al.*, eds. *Amazon peasant societies in a changing environment: political ecology, invisibility and modernity in the rainforest*. Doetinchem, Netherlands: Springer Science and Business Media Netherlands, 181–214.

Bryant, R. and Bailey, S., 1997. *Third world political ecology*. London: Routledge.

Burger, J., *et al.*, eds., 2001. *Protecting the commons: a framework for resource management in the Americas*. Washington D.C.: Island Press.

Calouro, A., 1995. *Caça de subsistência: sustentabilidade e padrões de uso entre seringueiros, riberinhos e não-riberinhos do Estado do Acre*. Thesis (MA), University of Brasília, Brazil.

Câmara, E., 1999. *Diagnóstico sócio–econômico e ecológico, Série Assentamento Sustentável São Salvador, Volume I*. Rio Branco, Acre, Brasil: PESACRE.

Chagnon, N., 1968. *Yanomamo: the fierce people*. New York: Holt, Rinehart and Winston.

Comunidade do Projeto de Desenvolvimento Sustentável São Salvador, 2003. *Plano de uso dos recursos naturais do Projeto de Desenvolvimento Sustentável São Salvador. Série Assentamento Sustentável São Salvador, Volume VI*. Rio Branco, Acre, Brazil: PESACRE.

Coomes, O. and Barham, B., 1994. The Amazon rubber boom: labor control, resistance, and failed plantation development revisited. *The Hispanic American Historical Review*, 74 (2), 231–257.

Coser, L., 1956. *The function of social conflict*. Glencoe, IL: The Free Press.

da Cunha, M.C. and Barbosa de Almeida, M., 2000. Indigenous people, traditional people, and conservation in the Amazon. *Daedalus*, 129 (2), 315–338.

da Cunha, M.C. and Barbosa de Almeida, M., Organizers, 2002. *Enciclopédia da floresta: o Alto Juruá: práticas e conhecimentos das populações*. São Paulo, Brazil: Companhia das Letras.

Davis, S., 1977. *Victims of the miracle: development and the Indians of Brazil*. Cambridge: Cambridge University Press.

Dean, W., 1987. *Brazil and the struggle for rubber: a study in environmental history*. Cambridge: Cambridge University Press.

Diegues, A., 1992. *The social dynamics of deforestation in the Brazilian Amazon: an overview*. Discussion Paper 36. United Nations Research Institute for Social Development.

Ellis, F., 1993. *Peasant economies: farm households and agrarian development*. Cambridge: Cambridge University Press.

Ellis, F., 2000. *Rural livelihoods and diversity in developing countries*. Oxford University Press, Oxford.

Fearnside, P., 1997. Environmental services as a strategy for sustainable development in rural Amazonia. *Ecological Economics*, 20 (1), 53–70.

Fragoso, J., Cunha dos Santos, M., and Nascimento, V., 2002. *A fauna silvestre e as práticas da caça. Série Assentamento Sustentável São Salvador, Volume III*. Rio Branco, Acre, Brazil: PESACRE.

Gibson, C., McKean, M., and Ostrom, E., 2000. *People and forests: communities, institutions, and governance*. Cambridge: MIT Press.

Godoy, R., Wilkie, D., and Franks, J., 1997. The effects of markets on neotropical deforestation: a comparative study of four Amerindian societies. *Current Anthropology*, 38 (5), 875–878.

Gudeman, S., 2001. *The anthropology of economy*. Malden, MA: Blackwell Publishers.

Gurven, M., *et al.*, 2000. Food transfers among Hiwi foragers of Venezuela: tests of reciprocity. *Human Ecology*, 28 (2), 171–217.

Hames, R. and McCabe, C., 2007. Meal sharing among the Ye'kwana. *Human Nature*, 18 (1), 1–21.

Harris, M., 2000. *Life on the Amazon: the anthropology of a Brazilian peasant village*. Oxford: Oxford University Press.

Harris, M., 2009. "Sempre ajeitando," (always adjusting): an Amazonian way of being in time. *In*: C. Adams, *et al.*, eds. *Amazon peasant societies in a changing environment: political ecology, invisibility and modernity in the rainforest*. Doetinchem, Netherlands: Springer Science and Business Media Netherlands, 69–91.

Harris, M. and Ross, E., 1987. *Food and evolution: toward a theory of human food habits.* Philadelphia: Temple University Press.

Hawkes, K., 1993. Why hunter-gatherers work. *Current Anthropology*, 34 (4), 341–361.

Hawkes, K., O'Connell, J., and Blurton Jones, N., 2001. Hadza meat sharing. *Evolution and Human Behavior*, 22 (2), 113–142.

Hecht, S., 2004. Invisible forests: the political ecology of forest resurgence in El Salvador. *In*: R. Peet and M. Watts, eds. *Liberation ecologies: environment, development, social movements.* New York: Routledge, 64–103.

Hecht, S. and Cockburn, A., 1990. *The fate of the forest: developers, destroyers and defenders of the Amazon.* New York: Harper Perennial.

Kainer, K., *et al.*, 2003. Experiments in forest-based development in western Amazonia. *Society and Natural Resources*, 16, 1–18.

Kaplan, H. and Hill, K., 1985. Food sharing among Ache Foragers: tests of explanatory hypotheses. *Current Anthropology*, 26 (2), 223–246.

Kelly, R., 1995. *The foraging spectrum: diversity in hunter-gatherer lifeways.* Washington D.C.: Smithsonian Institution Press.

Kensinger, K., 1983. On meat and hunting. *Current Anthropology*, 24 (1), 128–129.

Loker, W., 1993. The human ecology of cattle raising in the Peruvian Amazon: The view from the farm. *Human Organization*, 52, 14–24.

Lu, F., 2001. The common property regime of the Huaorani Indians of Ecuador: implications and challenges for conservation. *Human Ecology*, 29 (4), 425–447.

MacDonald, K., 2005. Global hunting grounds: power, scale and power in the negotiation for conservation. *Cultural Geographies*, 12 (3), 259–291.

Mauss, M., 1990. *The gift: the form and reason for exchange in archaic societies.* New York: W.W. Norton.

Meijaard, E., *et al.*, 2005. *Life after logging: reconciling wildlife conservation and production forestry in Indonesian Borneo.* Jakarta: Indonesia: CIFOR.

Melo, H., 2000. *O caucho e a seringueira, histórias da Amazonia, os mistérios da mata, os mistérios dos répteis e dos peixes, a expériencia do caçador, os mistérios dos pássaros e via sacra na Amazônia.* Rio Branco, Acre, Brazil: Fundação Elias Mansour, Governo do Estado do Acre.

Minnegal, M., 1997. Consumption and production: sharing and the social construction of use-value. *Current Anthropology*, 38 (1), 25–48.

Minzenberg, E., 2005. *Hunting and household in PDS São Salvador.* Thesis (PhD). University of Florida.

Nepstad, D., *et al.*, 2007. The costs and benefits of reducing carbon emissions from deforestation and forest degradation in the Brazilian Amazon. In: *United Nations Framework Convention on Climate Change Conference on the Parties*, Thirteenth Session, 3–5 December 2007 Bali, Indonesia. Woods Hole, MA: Woods Hole Research Center. Available from: http://www.whrc.org/resources/publications/pdf/WHRC_REDD_Amazon.pdf [Accessed 20 January 2011].

Neumann, R.P., 1992. Political ecology of wildlife conservation in the Mt. Meru area of northeast Tanzania. *Land Degradation and Rehabilitation*, 3 (2), 85–98.

Nugent, S., 1993. *Amazonian caboclo society: an essay on invisibility and peasant economy.* Oxford: Berg Publishers.

Nugent, S., 2009. Utopias and dystopias in the Amazonian social landscape. *In*: C. Adams, *et al.*, eds. *Amazon peasant societies in a changing environment:*

political ecology, invisibility and modernity in the rainforest. Doetinchem, Netherlands: Springer Science and Business Media Netherlands, 21–32.

Nygren, A., 2004. Contested lands and incompatible images: the political ecology of struggles over resources in Nicaragua's Indio-Maiz Reserve. *Society and Natural Resources*, 17, 189–205.

Oberschall, A., 1978. Theories of social conflict. *Annual Review of Sociology*, 4, 291–315.

Ostrom, E., 1990. *Governing the commons: the evolution of institutions for collective action*. Cambridge: Cambridge University Press.

Ostrom, E., *et al.*, eds. 2002. *The drama of the commons*. Washington D.C.: National Academy Press.

Pantoja, M.C., 2004. *Os Milton: cem anos de história nos seringais*. Recife, Brazil: Fundação Joaquim Nabuco, Editora Masangana.

Parker, E., ed., 1985. *The Amazon caboclo: historical and contemporary perspectives*. Department of Anthropology. Williamsburg, VA: College of William and Mary.

Patton, J., 2005. Meat sharing for coalitional support. *Evolution and Human Behavior*, 26 (2), 137–157.

Peet, R. and Watts, M., 1993. Introduction: development theory and environment in an age of market triumphalism. *Economic Geography*, 69 (3), 227–253.

Peluso, N., 1992. The Political Ecology of Extraction and Extractive Reserves in east Kalimantan, Indonesia. *Development and Change*, 23 (4), 49–74.

[PESACRE] Grupo de Pesquisa e Extensão em Sistemas Agroflorestais do Acre, 1998. *Sustainable settlement in the western Amazon: preliminary surveys of the São Salvador Resettlement Project in Acre, Brazil*. Proposal submitted to the W. Alton Jones Foundation. Rio Branco, Brazil: PESACRE.

Povinelli, E., 1992. Where we gana go now?: foraging practices and their meanings among the Belyuen Australian Aborigines. *Human Ecology*, 20 (2), 169–202.

Rao, M. and McGowan, P., 2002. Wild-meat use, food security, livelihoods, and conservation. *Conservation Biology*, 16 (3), 580–583.

Redford, K., 1992. The empty forest. *BioScience*, 42 (6), 412–422.

Redford, K. and Padoch, C., eds. 1992. *Conservation of neotropical forests: working from traditional resource use*. New York: Columbia University Press.

Robbins, P., 2004. *Political ecology: a critical introduction*. Malden, MA: Blackwell Publishing.

Robinson, J. and Bennett, E., 2000. *Hunting for sustainability in tropical forests*. New York: Columbia University Press.

Ross, E., 1978. The evolution of the Amazon peasantry. *Journal of Latin American Studies*, 10 (2), 193–218.

Roth, R.J., 2008. "Fixing" the forest: the spatiality of conservation conflict in Thailand. *Annals of the Association of American Geography*, 98 (2), 373–391.

Salisbury, D.S., 2002. *Geography in the jungle: investigating the utility of local knowledge for natural resource management in the western Amazon*. Unpublished thesis. University of Florida.

Salisbury, D.S. and Schmink, M., 2007. Cows versus rubber: changing livelihoods among Amazonian extractivists. *Geoforum*, 38 (6), 1233–1249.

Schmink, M. and Wood, C.H., 1987. The "political ecology" of Amazonia. *In*: P.D. Little, *et al.*, eds. *Lands and risk in the third world*. Boulder: Westview Press, 38–57.

Schmink, M. and Wood, C.H., 1992. *Contested frontiers in Amazonia*. New York: Columbia University Press.

Schwartzman, S., 1992. Land distribution and the social costs of frontier development in Brazil: social and historical context of extractive reserves. *Advances in Economic Botany*, 9, 51–66.

Scott, J., 1976. *The moral economy of the peasant*. New Haven, CT: Yale University Press.

Shaeff, G., 2002. *Avaliacão da comercialização. Série Assentamento Sustentável São Salvador. Volume IV.* Rio Branco, Acre, Brazil: PESACRE.

Simmel, G., 1955. *Conflict and the web of group affiliations*. Glencoe, IL: The Free Press.

Siskind, J., 1973. *To hunt in the morning*. New York: Oxford University Press.

State of Acre, 2009. *Plan for valuing forest assets: socio-environmental development policy of the State of Acre*. Rio Branco, Acre, Brazil: Environment State Secretary, State of Acre.

Stearman, A., 1989. Yuquí foragers in the Bolivian Amazon: subsistence strategies, prestige, and leadership in an acculturating society. *Journal of Anthropological Research*, 45 (2), 219–244.

Stonich, S., 1995. The environmental quality and social justice implications of shrimp mariculture development in Honduras. *Human Ecology*, 23 (2), 143–168.

Vadjunec, J.M., Schmink, M. and Gomes, C.V.A., forthcoming 2011. Rubber tapper citizens: emerging places, policies, and shifting identities in Acre, Brazil. *Journal of Cultural Geography*.

Wadley, R., Colfer, C., and Hood, I., 1997. Hunting primates and managing forests: the case of Iban forest farmers in Indonesian Borneo. *Human Ecology*, 25 (2), 243–271.

Wagley, C., 1953. *Amazon town: a study of man in the tropics*. London: Oxford University Press.

Wallace, R., 2004. *The effects of wealth and markets on rubber tapper use and knowledge of forest resources in Acre, Brazil*. Thesis (PhD). University of Florida.

Watts, M. and Peet, R., 2004. Liberating political ecology. *In*: R. Peet and M. Watts, eds. *Liberation ecologies: environment, development, social movements. 2nd Edition*. New York: Routledge, 3–47.

Wolff, C.S., 1999. *Mulheres da floresta: uma história Alto Juruá, Acre (1890–1945)*. São Paulo, Brazil: Editora Hucitec.

Yeh, E.T., 2000. Forest claims, conflicts and commodification: the political ecology of Tibetan mushroom-harvesting villages in Yunnan Province, China. *The China Quarterly*, 161, 212–226.

Zoneamento Ecológico-Econômico do Acre, 2000. *Indicativos para gestao territorial do Acre. Volume III*, Program Estadual de Zoneamento Ecológico-Econômico do Acre. Rio Branco, Acre, Brazil: Estado do Acre.

Traditional communities in the Brazilian Amazon and the emergence of new political identities: the struggle of the *quebradeiras de coco babaçu*—babassu breaker women

Noemi Porro[a], Iran Veiga[b] and Dalva Mota[c]

[a] *Amazonian Agricultures Graduate Program, Center for Agrarian Science and Rural Development, Federal University of Pará, UFPA;* [b] *Amazonian Agricultures Graduate Program at the Center for Agrarian Science and Rural Development, Federal University of Pará, UFPA;* [c] *Brazilian Research Enterprise for Agriculture and Livestock-EMBRAPA and Amazonian Agricultures Graduate Program at the Center for Agrarian Science and Rural Development, Federal University of Pará, UFPA*

Global environmental concerns have provided greater visibility to Amazonian traditional communities that have adopted new political identities to struggle for their livelihoods and territories. Through a case study of the *quebradeiras de coco babaçu*, babassu breaker women, this article offers a critical analysis of the challenges and opportunities involved in these processes. Through analyzing empirical evidence of their strategies of political representation, economic initiatives, a combination of productive and conservation concerns, and forms of accessing land and forest resources, this study addresses the risks of reproduction of relations of domination within their organizations, imposition of agendas by their allies and donors, and dilution of political capital due to the multiplication of social organizations. By confronting these risks, *quebradeiras'* organizations have managed to maintain a dynamic process of social learning. Embracing their internal diversity and sustaining a continuous dialogue between external actors and communities, they have been able to manage the tensions which emerged in their socio-political struggle. *Quebradeiras* have continuously reinvented their traditions, to strengthen new political identities and to bring concrete changes in the cultural geographies of their communities.

Introduction

Both natural and social landscapes in the Amazon are highly diverse and intrinsically entangled with an equally complex political landscape.

However, only recently did the dominant image of dense tropical rainforests begin to give space to conceptualizations of multiple Amazonian ecosystems, which range from daily flooded wetland forests to homogeneous palm forests, and from savannahs to submontane primary forests. Similarly, only in the past two decades has the social diversity of the Amazon begun to be visible to the general public, awakened by distinct social movements demanding effective political participation.[1] The internationally recognized struggle and martyrdom of the famous rubber tapper Chico Mendes (Schmink and Wood 1992) marked a momentous break from the simplistic myth that the Amazon consisted of pristine and homogeneous rainforests inhabited by indigenous groups who were seen as the only and eternal guardians of nature. Living images of social actors raising their voices to speak about specific public policies have largely displaced the neutral folkloric illustrations in many human geography textbooks and ethnologies that portrayed Amazonian residents almost as fixed parts of the ecosystem (IBGE 1956; see also Angotti-Salgueiro 2005; Almeida 2008).[2] Social movements made up of diverse cultural groups struggling for their ways of life not only have made themselves visible, but also brought gold miners, loggers, cattle ranchers, colonists, urban dwellers, entrepreneurs, environmentalists, priests and nuns into the public spotlight. The issues on the table are no longer only between nature and men, but also between a multitude of ecosystems and men and women from diverse geographical and historical backgrounds. Nature and populations have been politicized, and so has our understanding of Amazonian cultural geographies.

Growing environmental concerns, while providing fertile ground for greater visibility of traditional people and communities,[3] have also magnified the complexity of such political processes. Since the late 1980s, social visibility became an important strategy for traditional communities to affirm new political identities and to express their collective actions to the general public, aiming to achieve their rights in society. We define social visibility as both a strategy and a result of complex processes in which distinct social groups make their collective identity and social existence recognizable to society, in private and public domains (Scherer-Warren 2002; Esterci 2002). Political identity in this study means a continuous and dynamic construction related to recognition of differences between the collective selves and others, adopted by social groups aiming to negotiate and transform their condition in contexts of power differentials. The conceptualization of identity and its association with matters of political power is influenced by the works of Fredrik Barth (1969, 1981), Stuart Hall (1994, 1996) and Cardoso de Oliveira (2006).

When traditional communities began to make public demands in the late 1980s, the historically recognized organizations of representation and mobilization—from rural workers' unions at the municipal level to national organizations and confederations—were not able to embrace or express their distinct demands. Although remaining formally affiliated with these more widely recognized organizations, traditional communities have founded other, more specific organizations, launching a politics of their own and adding further complexity to ongoing processes of political negotiations. Although more than twenty years have passed since the first organizations of this kind were established, the degree of complexity of the political scene continues to challenge the academy, the government and leaders of social movements. At a governmental level, greater participation by traditional communities, in an attempt to incorporate local perspectives and practices into public policies, has not always achieved the desired results (Santilli 2005).

Researchers have shown how these movements and organizations emerge as a response by communities to dispossession of local resources or perceived threats to their lifeworlds (Bebbington *et al.* 2008b; Cronkleton *et al.* 2008). Their increasing interactions with national and transnational markets and the simultaneous transformations of existing systems of values and moral economies are also related to this response. With the support of external alliances and networks, including state agencies, these movements and organizations create new discourses based on their collective memory and shared experiences, impacting the communities' lifeworlds and productive practices (Perreault 2003). These new discourses question dominant concepts and models of development, and compete with them for the future of territories upon which local populations depend (Bebbington *et al.* 2008b). In spite of fostering institutional change, however, these movements and organizations often fail to bring about productive change that significantly improves the living conditions of local communities (Bebbington *et al.* 2008a). Moreover, they tend to reproduce social practices based on subordination of the local communities by external actors, and create so-called "clientelistic" relationships (Geffray 1995; Almeida 2000).

Clientelistic (or patron-client) relationships are conceptualized as relations resembling those established between a patron, who controls the access to material and/or symbolic resources, and clients, who may access these resources through personal relations with the patron, and not through a contract mediated by markets or third party rules. This relationship neither belongs to the public arena nor is ruled by public authorities, making it possible for the patron to define the relative value of the resource, creating a symbolic debt towards him, who "generously" permits access to resources. The relationship is sustained by the client's (and also the patron's) belief in the patron's generosity, such that the patron gains favor. This belief hides the harsh exploitation that may exist in these relationships, creating the idea of the "good patron." If circumstances change, the patron may be obliged to use coercive measures to maintain the relationship, becoming a "bad patron."

Through the case study of the *quebradeiras*' movement and its social organization MIQCB—Interstate Movement of the Babassu Breaker Women—this study explores two main questions. The first question analyzes the consequences of this discursive shift for their communities' practices, lifeworlds and social reproduction, and examines how it modifies the relationship between the movement's organizations and local communities. The second question seeks to understand the extent to which these organizations reproduce the clientelistic practices so prevalent in conventional patron-client relationships in Amazonia. This clientelistic pattern of social relations is so prevalent and strong in the lifeworlds of Amazonian peasants that it may be reproduced inside the organizations (unions, cooperatives, associations) created to challenge this very kind of exploitation, with the presidents, directors and sometimes technical staff playing the role of the "good patron."

We focus on how the *quebradeiras*' social movement and its grassroots organization, MIQCB, combine their ways of life based on the babassu palm forests (*Attalea speciosa*, also known as *Orbignya phalerata*) with their struggles to affirm their political identity using social visibility as a strategy. Since the early 1990s, the social movement of the *quebradeiras* has endured numerous challenges to negotiate local perspectives and the requirements to achieve social visibility. This article discusses how these women have managed to build social visibility while reaffirming their political identity, demonstrating the benefits of this strategy in spite of the continuous tensions it involves. The case study focuses on both challenges and opportunities of this social movement. It shows how their grassroots organization has faced challenges such as the fragmentation of social movements, and leaders' disconnection from grassroots politics. It also discusses how these women have explored opportunities to express their gender and ethnic-based identity and to prevent ruptures in the social networks.

Methodology

In this analysis, we use both primary data from fieldwork in *quebradeiras*' traditional communities in the Mearim Valley, in the state of Maranhão, Brazil and secondary data from the literature and related institutions' archives. Participant observation was carried out in several public events and rallies promoted by the *quebradeira* organizations throughout 2009–2010. Specific fieldwork for this article was carried out in two one-week periods, in the dry and rainy seasons of 2009.

Qualitative data at the household level was collected through 12 individual interviews in three villages in the Mearim Valley, and four focus group meetings with members of MIQCB—Interstate Movement of the Babassu Breaker Women, ASSEMA—Association in Settlement Areas in the State of Maranhão, COOPALJ—Cooperative of the Small Producers of Lago do Junco and COOPAESP—Cooperative of Small Producers of Esperantinópolis. Interviews with key informants were conducted at local and regional meetings and offices, and especially with the coordinators and technical advisors of *quebradeiras*' organizations, including MIQCB. In addition, the first author served as a research consultant in two negotiations involving agreements among the state, business enterprises and these organizations throughout 2009, allowing for additional direct and participant observation. The first process involved an intense and long negotiation between these organizations and a cosmetic enterprise to access babassu's genetic patrimony and associated traditional knowledge. The second involved issues of children's participation in traditional activities to produce babassu's goods for the "green" market. Both processes allowed for discussions on political identity and grassroots organization.

The movement of the *quebradeiras de coco babaçu*

Women who identify themselves as *quebradeiras de coco babaçu*—babassu breaker women—live in communities located within or near babassu palm forests in northern and northeastern Brazil (Figure 1). Their families make a living by gathering and breaking open the palm's fruits to extract oily kernels (Figure 2), in addition to slash-and-burn shifting cultivation (Figure 3). Babassu palms constitute an estimated 20 million hectares of continuous secondary growth forests, which have dominated vast tracts of the landscape since colonial settlers in the 17th century began to disturb primary forests (MIC/STI 1982; Anderson *et al.* 1991). The extractive activity of collecting and breaking open palm fruits to extract kernels has been connected to the international market at least since 1911, reaching a national production of 250,000 ton of kernels in the early 1980s (IBGE 1984). Babassu breaking is performed mostly by women and children,[4] while slash-and-burn shifting cultivation is carried out by the entire family and coordinated by men.

Babassu kernels are used for domestic consumption, but are mainly sold to the oil industry for the production of edible oil, margarine, soap and cosmetics. In traditional communities, the income from babassu may cover most of the costs of labor and inputs in small-scale family farming, until the harvest of their main staple, rice, obtained from slash-and-burn shifting cultivation on the so-called *roça*—agricultural fields planted in small patches of slashed palm forests. Men, women and children work in these fields, managing agricultural and extractive resources in the states of Maranhão, Pará, Tocantins and Piauí. Their territories,[5] within areas where babassu stands abound, include diverse phytogeographic regions, varying from moist forest zones in western and central Maranhão, southern Pará and northern Tocantins, through seasonally flooded low-lying wetlands in the Baixada region in the state of Maranhão, to

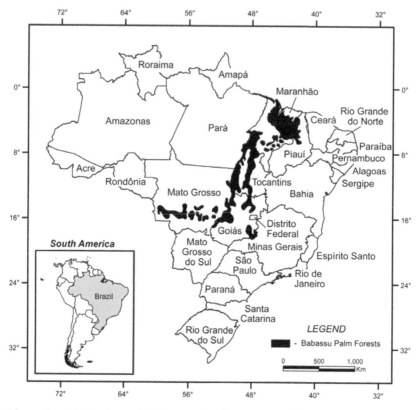

Figure 1. Distribution of babassu palm forests, according to the last governmental survey by LANDSAT (MIC/STI 1982), in the states of Maranhão, Piauí, Mato Grosso and Goiás—part of which belongs now to the state of Tocantins. Map adapted by the authors at the Remote Sensing Laboratory of EMBRAPA Eastern Amazon.

woodland, scrub and dry savannahs in the transition and the semiarid regions in the state of Piauí.

The peasantry constituting these traditional communities shares a common history of oppression and resistance. The diverse backgrounds of these communities involve interconnected histories of detribalization, enslavement, and forced migration. Currently, each *quebradeiras*' territory faces very specific social and political conditions. Some *quebradeiras* in traditional *quilombos*, communities of the descendents of runaway slaves, in the wetlands of Baixada Maranhense struggle against water buffalo ranchers encroaching on their lands. Meanwhile, other *quebradeira* colonists in official settlement projects in southeastern Pará face large industrial processing plants competing for babassu fruits to make charcoal for pig iron production. However, as a common fight, all *quebradeiras* struggle against getting trapped into what they call new

Figure 2. Woman breaking open babassu palm fruits to extract kernels. Maranhão, 2001. Photograph by Noemi M. Porro.

forms of *cativeiro,* captivity, meaning subordination to patrons or land-lords who grab their lands, and charge them for accessing babassu stands, or exploit their labor force. In traditional communities, they manage to remain free of subordination through *trabalho livre,* free labor, or the ability to work for oneself, without a boss, which integrates agriculture

Figure 3. Dos Santos family's rice field within babassu palm forests in Monte Alegre Village, Mearim Valley, Maranhão 2002. Photograph by Noemi M. Porro.

and extractive activities under family control, performed through specific gender and inter-generational relations (Porro 2002).

In their traditional communities, the *quebradeiras*' farming system is based on perceptions and practices of land and palm forests as common-use resources. Culturally defined rules prescribe a "first come, first served" norm, so that all must wait for the fruits to fall by themselves to be gathered, and nobody claims individual palm tree ownership. Research conducted since the 1970s (Mourão 1975; Mourão and Almeida 1974; Porro 1997; Carvalho 2000; Lago 2004; Figueiredo 2005) shows that, in each territory, although economic and environmental strategies may vary, babassu breaking is not just an economic activity to provide cash for agricultural families. Rather, it is a major symbol of peasant women's identity, and a material foundation for a distinct way of life. These studies demonstrate the centrality of their territories and *trabalho livre* for a successful *roça* supported by babassu breaking. These elements allow the autonomy and reproduction of the household, as a unit of production and consumption, according to their own culture, even within a context dominated by powerful landlords and a state governance structure which is mostly unfavorable to them.

In the late 1960s, federal and state laws and policies allowed cattle ranchers to expropriate traditional communities, privatizing public lands and legitimizing land grabbing (Almeida 1994). During the next thirty years, state and federal governments tacitly allowed the elimination of hundreds of traditional communities in favor of privately-owned estates, especially vast cattle ranches, resulting in land concentration and deforestation (Almeida 1995). During the late 1970s, grassroots initiatives began to erupt spontaneously against cattle ranchers who were slashing babassu palms and preventing women from gathering babassu, letting their cattle destroy peasants' rice fields, or planting pasture in forested areas hitherto managed by local communities. Throughout the 1980s, *quebradeiras*' groups supported by segments of the Catholic Church linked to Liberation Theology (Boff 1986), grassroots organizations, and NGOs began locally-based actions to make their struggles visible to the public. These mobilizations were located in six main regions: the Mearim Valley, the Baixada (Lowlands) region and in the surroundings of the municipality of Imperatriz, in the states of Maranhão; northern Piauí; southern Pará and northern Tocantins.

In 1991, 240 women from these four states rode in packed buses and convened in the capital of the state of Maranhão. For the first time they identified themselves as *quebradeiras de coco babaçu* in a large public event. Although the designations of *population of the babassu forests* and *quebradeira de coco babaçu* were well-known long before, it was mostly associated with victims condemned to a miserable job. In the 1950's, for example, officers of the Brazilian National Council of Economy reported to the president: "It is not a surprise for the population of the babassu forests to be the most miserable among the country. Semi-starved people, abandoned in an endless degradation, and in permanent nomadism" (Porro's translation) (National Council of Economy 1952, p. 9, cited in Almeida 2008). As a *quebradeira* stated:

> In the beginning, nobody would go to the city and say: I am a *quebradeira*. My holy lady, no! You were in the same old little dress, you were the same, but one would say: I am a farmer or I am a rural worker, because to say *quebradeira* meant you were broken, a have-not. . . . [Now] we know who we are, what we do. I know my fellows from here and from there, and we know we are in the same situation. (Porro's translation) (Querubina da Silva Santos, *quebradeira de coco babaçu*, 2004)

As they launched their movement, the *quebradeiras* proudly affirmed their identities, demanding the recognition of their rights. In their fourth meeting in 2001, with a general assembly of 240 members and a coordination of 12 leaders, an organization was formally founded. It was legally registered in 2002 as the *Associação do Movimento Interestadual das Quebradeiras de Coco Babaçu*, Association of the Interstate Movement of Babassu Breaker Women—MIQCB. The process of formalization and consolidation of MIQCB as a representative organization demanded a series of changes in leaders' and *quebradeiras'* practices. MIQCB was founded not only to represent all the diversity of the regional organizations and gain social visibility, but also to deal with new forms of accountability, to interact with advisors and technicians, and to negotiate with governments, donors, media and general society. This has demanded continual negotiation of differences and convergences, within their own group and between them and outsiders. In spite of the difficulties, the sixth interstate meeting in 2009, in São Luís, Maranhão, was a celebration of the successes and an evaluation of failures dealt with during these processes.

Risks and opportunities for new social movements

According to Scherer-Warren (2002), numerous and increasingly rapid transformations in the context of economic, technological and informational globalization have provoked changes in formats and subjectivities of social movements and the grassroots organizations representing them. In this scenario, a range of new issues emerges, surpassing conventional class struggles.

The *quebradeiras'* movement can be considered a "new" or "contemporary" social movement (Scherer-Warren 2002; Cohen 1985), which defines itself not only in terms of rural workers as a social class, but also as a grassroots movement of women who are members of traditional communities making a living from the babassu forest. The *quebradeiras'* movement emerged from a peasantry which integrates *quilombolas*, indigenous descendants, and migrants mostly from the northeastern Brazilian states of Maranhão, Paraíba, Piauí and Ceará—a diversity which confers unique cultural ties and political experiences. Therefore, although they remained as members of the rural unions, where gender and environmental issues were mostly overlooked in the 1980s and 1990s, *quebradeiras* needed a movement of their own to respond to threats to their traditions and aspirations.

These contemporary social movements "involve actors who have become aware of their capacity to create identities and of power relations involved in their social construction" (Cohen 1985, p. 694). They are self-limited in the sense that their aim is not to achieve radical social change, but to "reorganize relations between economy, state and society" (Cohen 1985, p. 670) in a more democratic and transparent way. To do so, their field of action is civil society, where new political identities can be created, and through this, changes can be made to conventional social relations in which those who form the base of the movement are engaged. The organizations created to represent these movements are not only a means to an end but are also an end in themselves, insofar as they reinforce and sustain these new political identities. This does not mean that concrete changes regarding either land or production are not important, but that a preceding issue for the movement is a change in discourse about itself and its territories, which helps pave the way for more sustainable changes in their access to resources and in their productive practices.

The *quebradeiras*' movement is thus quite different from feminist and environmentalist movements formed in the 1960s and 1970s in Western Europe and North America, to which the label "new social movement" was first applied. Although the *quebradeiras* struggle for greater gender equality, this issue is not their main concern. Even though the defense of the babassu palms is a major goal, they do not recognize themselves as environmentalists because their relation with the babassu forests is not primarily ruled by a conservation principle. Conservation is rather a consequence of a way of life, lived by men and women, whose traditions are rooted in and connected to specific resources and territories. Many of the features of the *quebradeiras*' movement can be explained by the political context originating in the 1988 Brazilian Constitution, which emphasized participation and diversity. The *quebradeiras*' movement is contemporaneous with the development apparatus that emerged world-wide in the 1980s and 1990s (Zaldivar 2005; Escobar 1995). This period was characterized by changes in actors, themes and philosophies of development. The end of the government's near monopoly over rural development created many new development actors: financial backers, international and national NGOs, agencies of cooperation, and local agents from diverse backgrounds and agendas, as well as new rural grassroots organizations. Consequently, the movement has also been shaped by the need to address neoliberal policies in Latin America.

Since the 1960s, this peasantry has been involved in the process of frontier expansion in the eastern Amazon, and has experienced ever greater market integration. These developments have brought complex social changes (through the *quebradeiras'* contact with different and hegemonic land tenure systems, as well as diverse social and cultural norms) and have had important impacts in the traditional communities, including threats to their access to resources and also to their lifeworlds (Bebbington *et al.* 2008b).

These threats, discussed in the introduction, are certainly one of the reasons for the emergence of the social movement and their organizations. The first response of these peasant communities was the creation—with the support of the Catholic Church and other external allies—of class-based organizations such as the rural workers' unions, in the 1970s and 1980s. The *quebradeiras'* organizations were created afterwards, in the 1990s, based on this first experience and, at least in the beginning, with the same members and network of external allies.

The emergence of this social movement is based on an alternative idea about the development of their communities in territories delineated by the use of babassu forests according to their own traditions. Such development implies free access to babassu as a common-use resource and free labor, and is directly oppositional to the dominant extensive cattle ranching development model. The translation of this alternative discourse as sustainable development was co-produced by the movement's organizations and external allies, such as NGOs, researchers and donors. These multi-scale networks (Perreault 2003) are very important in understanding the nature and management of *quebradeiras* organizations. Simultaneously, they entail the risk of imposing an external discourse on the movement.

New political identity and social visibility

Through these alliances, investments are made, locally and globally, to build images and discourses that link ways of life considered to be environmentally sustainable with resources and actions related to global environmental concerns. Building such images and discourses brings greater social visibility, which allows social groups such as the *quebra-deiras* to try to circumvent and eventually challenge local and regional power relations that are historically unfavorable to them. These alliances with the new development actors expanded their struggle for access to resources. What previously played out mainly in local and regional arenas was transformed into struggles in national and international arenas. This new situation made it possible for the *quebradeiras* to have access to different kinds of resources, natural, economic or symbolic, which were previously unattainable. It is important to note that these alliances also enable external actors to legitimize their action through the "reification of place and their insertion of the local into national or transnational politics" (Perreault 2003, p. 66). An example is the decade-long commercial relationship between COOPALJ, a *quebradeiras'* cooperative producing babassu oil in the Mearim Valley, and the Body Shop, a cosmetic industry selling around 34 "natural beauty" products with babassu as an ingredient. While *quebradeiras* accessed the green market to sustain their way of life and gained social visibility, the Body Shop validated its stated values of supporting local communities to protect the planet and defend human rights.

This broadened struggle for rights, which played out in national and global civil society arenas, has allowed somewhat greater inclusiveness and representation of the diverse realities and needs of *quebradeiras* within larger society. In the above example, the diverse realities and needs of both European and American consumers and *quebradeiras* are included in the commercialization of products promoting social and environmental justice concerns. To enter this relationship, the tradition of breaking open babassu fruits (usually for the middlemen and the patron) was not enough, but *quebradeiras* needed to renew the tradition to resist oppression and fight against power differentials. The new political identity thus created is not related to the *quebradeiras'* tradition understood as a fixed culture, but to a politically established tradition. "Tradition" was adopted by the social movement as a term expressing the goals of self-identification and autonomy as a political position (Almeida 2006a). The adoption of the term "traditional" with this meaning connects the *quebradeiras'* movement with concepts debated and affirmed during the elaboration of the International Labor Organization Convention 169 and the National Policy for Traditional Peoples and Communities (Decree 6040–2007). According to Almeida (2006b, pp. 65–67), "Traditional has nothing to do with immemorial . . . traditional is a way of being, a way of existing, a way of vindicating, of having a collective identity, which is a political experience as a group facing other groups and even the state itself." In the case of the *quebradeiras*, access to and control of agricultural and extractive resources through *trabalho livre*, defined here as the conditions in which labor is under the control of each family, is central to this struggle. Under *trabalho livre*, families are free from both landlords and bosses. The creation of this new political identity is strongly founded on these premises.

The relationship and the tensions between this politically reinvented identity and the actual grounded practices locally performed in the *quebradeiras'* community are central to our analysis. The potential for the *quebradeiras'* mobilization depends on practices based on grassroots solidarities. These solidarities are maintained in daily life by the relationships of the people of *roça* performing slash-and-burn, integrated with babassu breaking, through *trabalho livre*, free labor. At the same time, however, to mobilize effectively, the *quebradeiras* have to move beyond these local practices to create political practices performed under their new identity. To enter and be heard in a public arena, the *quebradeiras* have to present themselves as women defending palm forests through ecologically sound and socially just production, and have to embrace new solidarities with allies and donors.

The question to be addressed is how the movement and its main organization (MIQCB) negotiate this difficult connection, with its implicit risks of moving too far away from the diverse interests of their social base. According to the interviewed leaders and technical staff at MIQCB, this social base and their diverse interests are the main strength and reason for the existence of the organization. However, MIQCB and other *quebradeiras*' organizations are forced to seek double legitimacy: with their communities and with their external allies, each of them with their specific demands and interests. To these we could add the interests of the organizations themselves. This is a process that may take more time than fast-moving development processes allow. How to steer through this interface and to build an identity that makes sense to their social base? For example, as the coordinator of MIQCB stated, at the moment some within the movement might be feminists or environmentalists even without labeling themselves as such, but it would be wrong to adopt these labels because MIQCB as a whole is neither feminist nor environmentalist. In other words, the question facing the community involves how to initiate a real process of social learning (Hatchuel 1994; D'Incao and Roy 1995; Geslin 1999) instead of simply adopting ready-made categories such as "environmentalist" or "feminist" because they are pressured to do so.

There are three main challenges in this change process that we would like to address. The first one is the risk of reproducing, inside the movement and their organizations, the same unequal and exploitative clientelistic social relations that have characterized the history of this peasantry, such as their relationship with landlords and local merchants (Geffray 1995; Léna *et al.* 1996; Picard 1998; Albaladejo and Veiga 2002).

Zaldivar (2005), in his critique of the development apparatus, discusses a second risk: the multiplication of new political identities and social movements and their formal organizations, diluting the transformative potential of subordinate social groups and their class-based movements and organizations. He argues that the multiplication of small development projects can reinforce local patronage, establish non-democratic social structures around a leader, stimulate greater social differentiation within local communities, and create a state of "dependency" on the resources of these projects by peasants' political organizations, diluting their ability to operate effectively in the political arena.

Finally, when negotiating and implementing projects, when jointly promoting events, and even when *quebradeiras* participate in developing public policies, there is a risk that their allies will impose their own discourse on the movement. An example of this risk can be seen in representations of *roça*, their traditional slash-and-burn shifting cultivation, which is central to their livelihoods and lifeworlds, but does not fit into the environmentalists' perspectives. Slash-and-burn can be a very efficient method in integrated systems of production associated with indigenous and peasant ways of life in certain demographic and environmental settings. However, in the current context of increasing demographic densities in the babassu forests and in the face of climate change, environmentalist supporters and donors find it increasingly difficult to accept these cultivation practices, requiring intense negotiations between donors and recipients. *Quebradeiras* claim that the problem is not necessarily their *roças*, but the external variables that make *roça* unsustainable, such as land concentration and extensive removal of palms by cattle ranchers.

Through an analysis of four main issues – strategies of political representation, economic initiatives, forms of addressing agriculture and environment concerns, and access to land and forest resources – we will proceed with illustrations of how MIQCB has dealt with these challenges.

Strategies of political representation

In the 1990s, leaders of the rural workers' unions worried that the creation of MIQCB would cause a fragmentation of their class-based social movement. However, *quebradeiras* never left active membership in their rural workers' unions. As a matter of fact, *quebradeiras* are engaged in several organizations, from rural workers' unions to the National Council of Rubber Tappers, from Catholic organizations to the Landless Movement and political parties. They are engaged in party politics, and so far two of their leaders have been elected to municipal offices, based on their identity as *quebradeiras*. In fact, participation in different organizations with a network of multiple and diverse external actors is an important part of the essence of the *quebradeiras'* movement. For example, during our most recent fieldwork, we observed how *quebradeiras* were facing threats by Suzano Pulp and Cellulose, a top-ten entrepreneurial conglomerate intending to establish operations (eucalyptus plantations) around Imperatriz. To face such a powerful opponent, MIQCB was mobilizing and engaging *quebradeiras'* representatives and supporters from diverse organizations and institutions.

Throughout the years, although *quebradeiras* and MIQCB have adopted a wide range of strategies to achieve their goals, they have maintained the foundational principles of their tradition: a common struggle for a land without landlords; free access to the babassu palm as "the mother of the people;" and, in traditional communities, the ability to manage land and forests as common-use resources. The diverse formats and venues of representation and mobilization are an interesting strategy that works because *quebradeiras'* tradition is reflected in the foundations and practices of these formats. MIQCB has been successful in maintaining its mobilization and representation as legitimate so far because, although it is a formal organization, it maintains and articulates flexible, open, and locally defined strategies of representation.

Economic initiatives

Bebbington *et al.* (2008b) have discussed the difficulties for social movements to produce significant changes in the productive practices of their communities. The *quebradeiras'* organizations, since the beginning of their movement, tried to deal with institutional and economic changes at the same time. Economic initiatives have added legitimacy in MIQCB's representation, because *quebradeiras* conceive of effective representation and mobilization only if grounded in concrete and practical initiatives. By characterizing women's issues as central to their struggle, they aimed to change the economic condition of the whole family. In some places, they have been relatively successful in their economic pursuits. So far MIQCB has supported the formation of 26 collective production groups running grinding machines for babassu oil extraction at the community level, and processing soaps and mesocarp flour at the municipal level.[6] They are now organizing a cooperative in the Mearim and Imperatriz regions in Maranhão, and in northern Tocantins, southern Pará and Piauí. In the Mearim Valley, an affiliated grassroots organization, ASSEMA— Association in Settlement Areas in the State of Maranhão—has exported babassu oil for over 15 years. These *quebradeiras'* organizations sell their products to national and international consumers, both large and small. *Quebradeiras'* leaders, through MIQCB, also managed to have a seat on national committees in sectors of the Ministry of the Environment and Ministry of the Agrarian Development. As an example of the practical benefits of this political achievement, leaders and supporters managed, in 2009, to include babassu as one of the three undomesticated species named as "products of biodiversity" to be supported by a national minimum price policy established by Decree 79 of 1966 and Law 11,775 of 2008 (CONAB 2010).

All these initiatives aim to promote new productive practices, as well as greater interaction with national and international markets. These require *quebradeiras* to learn new institutional practices, including negotiations that allow their cooperatives and associations to survive in this market. This learning transition can be quite difficult, as shown by numerous failures of peasant cooperatives and of initiatives involving partnerships between local communities and enterprises. Throughout the 1990s several cooperatives and associations of *quebradeiras*, such as those in Imperatriz, Baixada, and São Luís Gonzaga, were closed due to internal difficulties, as well as incipient and oftentimes unsuitable economic incentives offered by the government to traditional peasant economies.

In addition, even those initiatives that managed to survive bring the risk, as discussed above, of resuming clientelistic practices, common between patron and clients among Amazonian communities. However, because the *quebradeiras'* movement has maintained some of the grounded family and community rules in their collective initiatives, conflicts that arise have been managed adeptly. By linking traditional values – honor, shame, family, respect – to new initiatives, the *quebradeira's* movement has largely avoided the potential risks of clientelistic labor relations, and instead secured *trabalho livre* practices in their communities. There are indeed situations in which their own leaders and technicians tend to act as new patrons, even though they may serve as "good patrons." However, these situations do not last, because of social control by communities based on the values mentioned above. The social control, exercised by women at the local productive units, is made possible because these economic initiatives are executed and directed at grassroots levels.

Forms of addressing environment and productive concerns

In the present global context of intense focus on environmental concerns, exacerbated by concerns about climate change, another relevant issue is the divergent perceptions and practices regarding agriculture held by *quebradeiras*, donors, governments and NGOs. Slash-and-burn shifting cultivation, *roça*, is one of the causes of deforestation, but it is also an important component of *quebradeira* families' livelihoods. In images and discourses representing *quebradeiras*, *roça* is often minimized or omitted by supporters driven by conservationist goals. However, *roça* is a fundamental component of their identities, and a social practice performed through cultural rules that constantly revitalize intra- and inter-family social relations within communities.

MIQCB has kept the discourse and practice of *roça* as fundamental to *quebradeira* livelihoods in its projects, although doing so increases the risk of misunderstanding between donors and recipients. In fact, several projects implemented by MIQCB are based on environmentalist assumptions which do not always coincide with *quebradeira* daily concerns and local cultural practices. Leaders and technical staff have so far succeeded in managing these differences, seeking constant dialogue about them with donors. This dialogue is productive only when donors are open to learning about the local realities of *quebradeiras*. In these instances, both donors and recipients engage in discussions searching for alternatives acceptable to both groups.

On the one hand, the MIQCB is aware of the critical situation of the *quebradeiras*, in their constrained contexts of scarce resources. While promoting debates about these different perspectives within communities, the MIQCB also participates actively in public arenas where these problems are negotiated and addressed, for example, in the committees within the Ministry of the Environment and the Ministry of Agrarian Development. On the other hand, the centrality of slash-and-burn shifting cultivation under family control does not mean that it is immutable. The MIQCB is researching alternative forms of practicing small-scale family farming, such as agroforestry gardens and planting on unburned *roça* using green manure. The *quebradeiras* and their supporters are not against changes in slash-and-burn shifting cultivation, but they want to be in control of the nature, rhythm and intensity of these changes as a group, fostering a social learning process.

Access to land and forest resources

Quebradeiras possess a vast range of formal and informal relations with the land as well as with babassu palm forests and other natural resources. *Quebradeiras* may live as smallholders with legal title, as landless urban periphery residents gathering babassu in neighboring cattle ranches, as agrarian reform beneficiaries in extractive reserves, or as colonists in individually owned and managed 100-hectare plots. Since the mid-1980s and throughout the 1990s, several communities managed to attain legal recognition of their traditionally occupied lands. An interesting feature of the agrarian reform in several traditional communities was that lands were maintained as a common resource, disregarding government attempts to divide them into individual plots. In these cases, babassu palm forests continued to be common-use resources as well. However, in contrast, hundreds of *quebradeiras* lost their territories and became landless or remained on lands without any legal recognition of their traditional tenure. Also, in conventional schemes of agrarian reform areas, especially in new frontier regions in the states of Pará and Tocantins, land was in fact divided into individual plots. In these situations, social rules that maintain the logic of land as a common resource change and internal conflicts may emerge. Whereas landless women continue to consider that they have the right to gather babassu wherever the palms grow, families, especially men, owning individual plots may consider the palms as their exclusive property.

This situation shows how the idea of a common resource is at the core of *quebradeiras* political struggles as women whose livelihood depends on *roça* and the babassu palm as "the mother of the people." Even where it is no longer possible to apply this tradition towards the land, *quebradeiras* still insist on sharing babassu palms as common resources in their traditional communities. In fact, although agrarian reform has remained an important demand, another historical claim of the *quebradeiras* became, in 2003, the main banner of MIQCB. More specifically, the "Free Babassu Law" is a grassroots rule and local practice demanding free access to babassu palms, even on private property (Shiraishi 2006). Although fully aware of the legislation privileging conventional private property, and policies favoring privatization of public resources, *quebradeiras* have insisted on their understanding of babassu palm forests as a free access and common-use resource. Therefore, even when land is legally privatized, *quebradeiras* continue to consider palms as a common resource. In the words of one *quebradeira*, "nobody planted it; nobody watered it; it is for the people." From 2003 to 2010, seventeen municipal councils in the eastern Brazilian Amazon have approved "Free Babassu" laws because of *quebradeira* lobbying efforts. Since 2004, four proposals have been submitted to the Brazilian Congress petitioning for a federal law to prohibit removal of babassu palms and allowing free access to the fruits by members of traditional communities, regardless of their legal landholding situation.

MIQCB, although very busy with economic initiatives and involved in issues not covered by the unions, is nevertheless aware of the risks of distancing itself from the struggle for land, a major banner of the rural workers' unions. They know that the integrity of their ways of life depends on their territories, and that the "Free Babassu" laws are but a relevant means to the ultimate end of regaining them. For some groups this realization dates to the 1980s. When asked whether the conflict began with a struggle for land, a *quebradeira* from the social movements in the Mearim Valley answered: "In the beginning, the fight was because of the babassu; we needed the babassu-nuts to survive, we live on babassu. So, we were forced to react, to no longer fear the cattle rancher . . . As women began to react and men to support them, we started to fight for the babassu-nut and later for the land" (quoted in Lago 2004, p. 36). For some other groups, this realization is just emerging. The "Free Babassu" law is also a means to maintain collective social learning in spite of all these diverse situations.

At each step in their development, the *quebradeiras* have learned to gradually establish limits to what is and what is not negotiable. The foundations of their lifeworlds – free labor, free access to babassu palms, and, at least for groups in traditional communities, the common use of land – are not negotiable in principle, but how to stick to them while maintaining *roça* integrated with babassu breaking has demanded transformations in their traditions.

Conclusion

The case study of the *quebradeiras de coco babaçu* working to defend their territories breaks with the image of Amazonia as a homogeneous place. Women from a distinct, and often invisible cultural group, struggling for their livelihood in a mono-species palm forest, seek social visibility along with the indigenous people who guard the diverse and lush tropical jungle. Certainly, the *quebradeiras'* movement and MIQCB have much to celebrate in terms of advances in access to land, forest resources and economic initiatives. Above all, the movement's symbolic advances are impressive and outstanding in the cultural Amazonian landscape. They have transformed women, previously relegated to an activity disdained even by their own peers, whose product was undervalued in the market, into protagonists of one of the most visible women's movements based on a peasant way of life in Brazil. This social visibility helped the *quebradeiras* to create a new political identity, and has provided opportunities for concrete initiatives on the ground.

However, throughout the process, permanent tensions exist between images and discourses built to establish such a visibility, and the daily reality lived by *quebradeiras* and represented by MIQCB. While these tensions shape the social interactions within the group and between them and external actors, they have also fostered social learning. Surely such images and discourses created in and for domains mostly controlled by outsiders are influencing local realities. However, as much as these external domains promote changes in local communities, the grassroots initiatives of the *quebradeiras* also have influence in changing these external domains. As *quebradeiras* very skillfully enmesh their local realities in the process of building images and discourses, tensions do not always end in ruptures. In the final assessment, images and discourses are never only for external consumption. In these tense interactions, *quebradeiras* have so far successfully established a visible space for social learning and affirmation of their political identity.

It is important to analyze how, within the political and economic contexts in which they are inserted, *quebradeiras* have overcome these tensions and avoid clientelistic practices. We should highlight the importance of their agency in accepting and managing the internal diversity of their movement. All social movements are of course diverse, but MIQCB has managed to put together and, in spite of the high costs to its leaders and members, to maintain a cohesive movement run by people from very different contexts and interests. Embracing this diversity has been a key to the movement's achievements. The practice of dealing with difficulties to coordinate socially and geographically diverse groups (Bebbington *et al.* 2008b) becomes a strength rather than a weakness, because it generates powerful openness both to internal differences and to connections with diverse external actors.

This openness among members also influences the internal governance of the movement and its organizations, particularly the MIQCB. It requires that *quebradeiras* from different regions and social contexts be represented in the coordination of the organization. As registered in MIQCB's reports, since their second general meeting in 1993, there is a significant rotation of members on the board of coordinators, while founder leaders still remain active in the movement. This helps to prevent leadership concentration and perpetuation, traits which are commonly known to weaken the long-term sustainability and democratic aims of community organizations. It also encourages the organization to be open to very different ideas, trajectories, demands and external contacts.

In addition, the diversity of *quebradeiras'* demands and social situation foster MIQCB's necessity to balance pragmatism and constant reflection in defining objectives and negotiation strategies with allies and antagonists. Based on general goals shared by most *quebradeiras* including free access to babassu palms as a common resource, rights to traditional lands, gender equality, and free labor, MIQCB has had to exercise constant flexibility to be able to address its members' different demands and interests (MIQCB 2009). This is reflected in the wide range of issues dealt with by the organization, which is in itself a source of social learning.

Finally, the internal diversity of the *quebradeiras* has led to the formation of an equally diverse range of allies, collaborators and support networks. This large range of external collaborators and contacts, from large companies to governmental agencies, from Catholic organizations to feminist and environmentalist institutions, is related to their decision to interact in the world of development. Dealing with this broad array of actors provides continuous learning about markets and the public arena, as well as the circulation of new ideas. More importantly, it prevents MIQCB from becoming too dependent, both materially and symbolically, on a small group of allies and donors. The risk of reproducing clientelistic relations with allies, as discussed earlier, is smaller when there are greater possibilities of external relations. This diverse network of allies and contacts enables MIQCB to act at different scales, from local to international, and to create and support affiliated organizations to act through different means, from community associations to interstate cooperatives. This, in turn, reinforces the internal and external openness of the organization and the movement.

Acknowledgements

The authors would like to thank each and every *quebradeira de coco babaçu* who patiently shared lessons about their ways of life and social movement.

Notes

1. According to Scherer-Warren (2009, p. 1): "there is a social movement when a collective action generates a principle of identity as a group, defining opponents or antagonists who work against the thorough existence of this identity and identification, and acting on behalf of a process of change, either social, cultural or systemic." While social movements are non-hierarchical and fluid, social organizations or grassroots organizations are the hierarchical structures formalized to represent and mobilize them, as an agent in the interlocution with other sectors in society.
2. Consider, for example, the illustrations by Percy Lau, in *Tipos e Aspectos do Brasil* for the *Revista Brasileira de Geografia* (IBGE 1956). According to Almeida (2008), Lau's illustrations were used in textbooks for geography classes in schools in Brazil from the 1950s to the 1970s.
3. In Brazilian legislation, traditional people and communities are defined as: "culturally differentiated groups who recognize themselves as such, who hold their own forms of social organization, who occupy and use territories and natural resources as a condition for their cultural, social, religious, ancestral and economic reproduction, utilizing knowledge, innovations and practices generated and transmitted through tradition" (Decree 6040 of February 7, 2007).
4. According to the IBGE (2008a) census for agriculture and livestock, there were 46,706 landholdings producing babassu kernels. Women's grassroots organizations estimate 400,000 babassu breakers (Almeida 2006c, p. 49). At any rate, they are the largest group of traditional communities extracting forest products, and the second in terms of value of production. The value of the 114,847 tons of kernels produced in 2007 was R\$ 113,268,000 or approximately U\$ 56,634,000 (IBGE 2008b, 2008c).
5. Territories are defined here as the units made up of the physical land and associated natural resources, intrinsically linked to all the socially constructed attributes that assure their belonging to a people and to a way of life collectively performed in a specific place, throughout their common history and political struggles.
6. Babassu fruits have a peel (epicarp) covering a starchy layer (mesocarp), which envelops a hard shell (endocarp) holding oily kernels.

References

Albaladejo, C. and Veiga, I., 2002. Introdução: organizações sociais e saberes locais frente à ação de desenvolvimento na direção de um território cidadão. *Agricultura Familiar Pesquisa, Formação e Desenvolvimento*, 1 (3), 1–12.

Almeida, A.W.B., 1994. Transformações agrárias e conflitos sociais nas áreas de ocorrência do *babaçu*. *CEDE*, 3 (IV), 43–60.

Almeida, A.W.B., 1995. *Quebradeiras de coco babaçu: identidade e mobilização. Movimento interestadual das quebradeiras de coco babaçu*. São Luís: Estação Publicações Ltda.

Almeida, A.W.B., 2000. *O GTA face aos objetivos do BRA/96/012. Avaliação independente do projeto de fortalecimento institucional do Grupo de Trabalho*

Amazônico–GTA, BRA/96/012. Brasília: unpublished independent report to GTZ.

Almeida, A.W.B., 2006a. Arqueologia da tradição. *In*: J. Shiraishi, ed. *Leis do Babaçu Livre: práticas jurídicas das quebradeiras de coco babaçu e normas correlatas.* Manaus, AM: Fundação Ford, 7–12.

Almeida, A.W.B., 2006b. Identidades, territórios e movimentos sociais na Pan-Amazônia. *In*: R. Marin and A.W.B. Almeida, eds. *Populações tradicionais: questões de terra na Pan-Amazônia.* Belém, PA: UNAMAZ, 60–70.

Almeida, A.W.B., 2006c. *Terras de quilombo, terras indígenas, "babaçuais livres", "castanhais do povo", faxinais e fundos de pasto: terras tradicionalmente ocupadas.* Manaus, AM: PPGSCA/UFAM.

Almeida, A.W.B., 2008. *Antropologia dos archivos da Amazônia.* Rio de Janeiro: Casa 8/ Fundação Universidade do Amazonas.

Anderson, A.B., May, P., and Balick, M.J., 1991. *The subsidy from nature: palm forests, peasantry, and development on an Amazon frontier.* New York: Columbia University Press.

Angotti-Salgueiro, H., 2005. A construção de representações nacionais: os desenhos de Percy Lau na Revista Brasileira de Geografia e outras "visões iconográficas" do Brasil moderno. *Anais do Museu Paulista* 13(2) 21–72. Jul–dez 2005.

Barth, F., 1969. *Ethnic groups and boundaries: the social organization of cultural difference.* Prospect Heights, Ill: Waveland Press.

Barth, F., 1981. *Process and form in social life.* Boston: Routledge & Kegan Paul.

Bebbington, A., Abramovay, R., and Chiriboga, M., 2008a. Social movements and the dynamics of rural territorial development in Latin America. *World Development*, 36 (12), 2874–2887.

Bebbington, A., *et al.*, 2008b. Mining and social movements: struggles over livelihood and rural territorial development in the Andes. *World Development*, 36 (12), 2888–2905.

Boff, L., 1986. *Church, charism and power: liberation theology and the institutional church.* New York: Crossroad.

Carvalho, C.M., 2000. *Agricultura extrativismo e garimpo na lógica camponesa.* Thesis (Master). Universidade Federal do Maranhão, São Luis.

Cohen, J.L., 1985. Strategy or identity: new theoretical paradigms and contemporary social movements. *Social Research*, 52 (4), 663–716.

CONAB—Companhia Nacional de Abastecimento, 2010. Plano operacional anual de apoio à comercialização de produtos da sociobiodiversidade safra 2010/11. Brasília: Ministério da Agricultura, Pecuária e Abastecimento. Available from: http://www.conab.gov.br/OlalaCMS/uploads/arquivos/b0d8e992de2cb83cecdecca53984d3ce..pdf [Accessed 19 January 2010].

Cronkleton, P., *et al.*, 2008. Environmental governance and the emergence of forest-based social movements. *Occasional Paper* 49. Bogor: CIFOR.

D'Incao, M.C and Roy, G., 1995. *Nós, cidadão: aprendendo e ensinando a democracia.* Rio de Janeiro: Paz e Terra.

Escobar, A., 1995. *Encountering development: the making and unmaking of the Third World.* Princeton: Princeton University Press.

Esterci, N., 2002. Diversidade sociocultural e políticas ambientais na Amazônia Brasileira: o cenário contemporâneo. *Boletim Rede Amazônia*, 1 (1), 3–5.

Figueiredo, L.D., 2005. *Do espaço doméstico ao espaço público—lutas das quebradeiras de coco babaçu no Maranhão.* Thesis (Master). Universidade Federal do Pará, Belém.

Geffray, C., 1995. *Chroniques de la servitude en Amazonie brésilienne.* Paris: Karthala.

Geslin, P., 1999. *L'apprentissage des mondes. Une anthropologie appliquée aux transferts de technologies.* Toulouse/Paris: Éditions de la Maison des sciences de l'homme.

Hall, S., 1994. Cultural identity and diaspora. *In:* P. Williams and L. Chrisman, eds. *Colonial discourse and post-colonial theory.* New York, NY: Columbia University Press, 392–403.

Hall, S., 1996. Introduction: who needs identity? *In:* S. Hall and P. Du Gay, eds. *Questions of cultural identity.* London: Sage Publications, 1–17.

Hatchuel, A., 1994. Apprentissages collectifs et activités de conception. *Revue Française de Gestion*, juillet-août, 109–120.

IBGE—Instituto Brasileiro de Geografia e Estatísitca, 1984. Censo Agropecuário 1980. Rio de Janeiro: FIBGE.

IBGE—Instituto Brasileiro de Geografia e Estatísitca, 2008a. Censo Agropecuário 2006. Tabela 816. Available from: http://www.sidra.ibge.gov.br/bda/tabela/listabl.asp?z=t&o=18&i=P&c=816 [Accessed 11 November 2010].

IBGE—Instituto Brasileiro de Geografia e Estatística, 2008b. Banco de Dados SIDRA. Quantidade produzida e valor do extrativismo vegetal por tipo de produto extrativo. Table 289. Available from: http://www.sidra.ibge.gov.br/bda/tabela/protabl.asp?c=289&z=t&o=18&i=P [Accessed 11 November 2010].

IBGE—Instituto Brasileiro de Geografia e Estatística, 2008c. Banco de Dados SIDRA. Quantidade produzida e valor do extrativismo vegetal por tipo de produto extrativo. Table 290. Available from: http://www.sidra.ibge.gov.br/bda/tabela/protabl.asp?c=290&z=t&o=18&i=P [Accessed 11 November 2010].

IBGE—Instituto Brasileiro de Geografia e Estatística and Conselho Nacional de Geografia, 1956. *TIPOS e aspectos do Brasil (excertos da Revista Brasileira de Geografia).* Ilustrações de Percy Lau, 6th ed. Rio de Janeiro: IBGE/Conselho Nacional de Geografia.

Lago, R.T., 2004. *Babaçu livre e roças orgânicas—a luta das quebradeiras de coco babaçu do Maranhão em defesa dos babaçuais e em busca de formas alternativas de gestão dos recursos naturais.* Thesis (Master). Universidade Federal do Pará, Belém.

Léna, P., Geffray, C. and Araújo, R. 1996. *L'oppression paternaliste au Brésil. Lusotopieo*, 105–108.

MIC/STI—Ministério da Indústria e do Comércio, Secretaria da Tecnologia Industrial, 1982. *Mapeamento e levantamento do potencial das ocorrências de babaçuais.* Map of Brazil without scale, detailed maps in 1:1.000.000. Brasília: MIC/STI.

MIQCB—Movimento Interestadual das Quebradeiras de Coco Babaçu, 2009. Nos babaçuais há conhecimentos tradicionais. Report of the 6th interstate meeting of the *quebradeiras de coco babaçu.* São Luís: Gráfica Santa Clara.

Mourão, L., 1975. *O Pão da Terra.* Thesis (Master). Universidade Federal do Rio de Janeiro, Rio de Janeiro.

Mourão, L. and Almeida, A.W.B., 1974. *Questões Agrárias no Maranhão Contemporâneo.* Unpublished Manuscript. São Luís.

145

Oliveira, R.C., 2006. *Caminhos da Identidade: ensaios sobre etnicidade e multiculturalismo*. São Paulo: Editora Unesp; Brasília: Paralelo 15.

Perreault, T., 2003. Changing places: transnational networks, ethnic politics, and community development in the Ecuadorian Amazon. *Political Geography*, 22, 61–88.

Picard, J., 1998. *Amazonie brésilienne: les marchands de rêves. Occupations de terres, rapports sociaux et développement*. Paris: L'Harmattan.

Porro, N.M., 1997. *Changes in peasant perceptions of development and conservation*. Thesis (Master). University of Florida, Gainesville.

Porro, N.M., 2002. *Rupture and resistance: gender relations and life trajectories in the babaçu palm forests of Brazil*. Thesis (PhD). University of Florida, Gainesville.

Santilli, J., 2005. *Socioambientalismo e novos direitos: proteção jurídica à diversidade biológica e cultural*. São Paulo: Peirópolis.

Scherer-Warren, I., 2002. A atualidade dos movimentos sociais rurais na nova ordem mundial. *In*: I. Scherer-Warren and J.M.C. Ferreira, eds. *Transformações Sociais e Dilemas da Globalização: Um Diálogo entre Brasil/Portugal*. São Paulo, SP: Cortez Editora, 243–257.

Scherer-Warren, I., 2009. Movimentos sociais na América Latina: revisitando as teorias. *In*: Anais do Congresso Brasileiro de sociologia, 14. Rio de Janeiro, RJ: SBS. Available from: http://www.npms.ufsc.br/wpapers.php [Accessed 1 August 2010].

Schmink, M. and Wood, C., 1992. *Contested frontiers in Amazônia*. New York: Columbia University Press.

Shiraishi, J., 2006. *Leis do babaçu livre: práticas jurídicas das quebradeiras de coco babaçu e normas correlatas*. Manaus: UFAM/Fundação Ford.

Zaldivar, V.B.S., 2005. *Capital social y etnodesarrollo en los Andes*. Quito: Centro Andino de Acción Popular.

Transboundary political ecology in Amazonia: history, culture, and conflicts of the borderland Asháninka

David S. Salisbury[a], José Borgo López[b] and Jorge W. Vela Alvarado[c]

[a]University of Richmond, Department of Geography and the Environment, Richmond, VA, USA; [b]Dirección Forestal y de Fauna Silvestre, Atalaya, Ucayali, Perú; [c]Universidad Nacional de Ucayali, Centro de Investigación de Fronteras Amazónicas, Pucallpa, Ucayali, Perú

International boundaries in the lowland Amazon forest were historically drawn according to the scramble for natural resources. This paper uses a case study from the Peruvian and Brazilian border and the Ucayali and Juruá watersheds to understand the political ecology of a border process from contact to 2004. Results demonstrate how global resource demand and ecological gradients drove boundary formation and the relocation of indigenous labor to the borderlands. Forgotten in the forest after the fall of rubber prices, the borderland Asháninka emerged to challenge loggers incited by the global demand for high grade timber. The transboundary impacts of this resource boom highlight discrepancies between the Brazilian and Peruvian Asháninka's ability to mobilize power. A transboundary political ecology framework is necessary to grasp the heterogeneity and dynamism of natural resource management along boundaries and borderlands forged and tempered by historical resource booms.

Introduction

In the remote southwestern borderlands of Amazonia shared by Peru and Brazil dwell a formerly invisible indigenous people, the borderland Asháninka, increasingly threatened by the global demand for high grade timber. This demand resembles the rubber boom that brought the Asháninka to these nascent borderlands one hundred years ago. To make sense of the borderland Asháninka's past and present this paper introduces the historical dimensions of transboundary political ecology through a case study of the history, conflicts, and peoples of an increasingly important, if little known, corner of Amazonia. Transboundary political ecology builds on the historical political ecology framework proposed by Offen (2004) as a field-informed analysis of human-environment relations in the past with significance for improving conservation and environmental/social justice today. Thus, the article's detailed results, from both archival and ethnographic methods, not only serve as documentation for the marginalized groups described within but also inform the historical and cultural ecology elements of cultural geography through the rural nature, historical approach, and indigenous focus of the article.

Fieldwork along the international boundary between the Ucayali and Juruá Rivers revealed complex cultural geographies and dynamic identities as the borderland Asháninka people struggled against the incursions of illegal loggers. Two related and neighboring Asháninka groups marshaled disparate amounts of power in the face of the invaders: the borderland Asháninka in Brazil, a titled, empowered, and globally recognized people; and those in Peru, an untitled, marginalized, and invisible people. Following fieldwork, archival research traced a faint but complex trail woven into the political geography and ecology of the Amazon borderlands. This trail connects with multiple themes within this special issue: shifting cultural landscapes, hidden histories, heterogeneity, and political economy, to name a few. Before sharing this trail, the research is mapped in the literatures of cultural/political ecology, political geography, and the Asháninka people.

Cultural and political ecology

The varied concepts and concerns loosely grouped under the label of cultural and political ecology lead scholars to increasingly see this as a vibrant and wide-ranging approach rather than a narrow subdiscipline (Zimmerer and Bassett 2003; Robbins 2004; Neumann 2005). While researchers continue to debate the approach's bias towards politics (Peet and Watts 2004; Walker 2007), ecology (Walker 2005; Walters and Vayda 2009), or a particular scale (Brown and Purcell 2005; Neumann 2009), this research combines cultural ecology's nuanced understanding of culture-nature relationships with political ecology's focus on contextualizing resource management and its impacts within political economies at multiple scales (Hecht 2004).

Some of political ecology's earliest efforts were situated in the region of Amazonia (Hecht 1985; Schmink and Wood 1987; Chapman 1989) with particularly well-known works focused on the differential access to power (Schmink and Wood 1992) and resistance (Hecht and Cockburn 1990) of stakeholders facing land use change under authoritarian policies. More recent Amazonian political ecology scholarship also investigates the policy relationships related to land use change (Aldrich *et al.* 2006; Vadjunec and Rocheleau 2009; Walker *et al.* 2009) or territorial conflict (Little 2001; Porro 2005; Simmons *et al.* 2007).

Following Offen (2004) this paper constructs a transboundary political ecology framework useful for investigating policy, conflict, and land-use change in a borderland context through Hecht's (2004) three historical political ecology themes: the mythic empty Amazon, the scientific Amazon, and the production of landscape. These themes encompass not only an expansive political ecology of Amazonia, but, as used here, also shed light on the complex cultural and historical geographies embedded in similarly diverse, dynamic, and distant borderland regions.

The first theme, the empty Amazon landscape or tropical *tabula rasa*, provides an imagined blank slate or empty forest for the resource-hungry colonial powers of the past or the insatiable global marketplace of today to design the nature and future of the Amazon basin. The very political boundaries of Amazonia were carved into the basin via political struggles over natural resources in the poorly mapped Amazon interior. Interestingly, current borderland conflicts imitate these colonial clashes.

Hecht's (2004, p. 47) scientific Amazon approach refers to not only the tension between a climax "forest primeval" and the dynamic tropical forest of today, but also "that environments have an effect on peoples and societies, not in the ways understood in environmental determinism, but rather as mediated by its symbolic and cultural meanings, as well as resource possibilities." This research adopts a similar stance by mediating analysis of the environment through resource possibilities, investigating the power and importance of watershed divides and biogeographical gradients to the harvesting of natural resources, the formation of political boundaries, and the movement of peoples.

Within the final theme, the production of landscapes, Hecht (2004) underscores the importance of hidden social histories, landscapes, and development impacts. This paper shows, for example, how the borderland Asháninka of Peru both arrive in and emerge from an unmapped and "empty" borderscape, through the ebb and flow of past and present resource booms.

Political geography

Robbins (2003) argues for the theoretical potential of melding political geography with political ecology to produce conceptually advanced explanations of complex human-environment interactions. Rather than attempt to join these broad approaches in their entirety, this research bridges political geography's expertise on boundaries and borderlands with the Amazonian political ecology described above. In linking political ecology and political geography through borderland study, this research addresses three key topics identified by boundary scholars: a more detailed multi-scalar examination of the transboundary environment (Newman and Paasi 1998), a need to analyze borders as dynamic processes (Newman 2006), and more border analysis that incorporates the local scale (Newman 2006).

As applied to the Amazon borderlands, these boundary topics fit within Hecht's (2004) three themes and Offen's (2004) historical political ecology framework for the following three reasons: (1) multi-scalar examination, and particularly local analysis, are critical to uncovering the hidden cultural geographies, political ecologies, and social histories of these Amazon borderlands and the Asháninka people; (2) historical analysis makes it possible to trace the dynamic political ecology of border/borderland processes from contact to the 21st century; and (3) the tropical *tabula rasa* can only be accurately described with the pointillist approach of local scale analysis. Despite the synergy between these boundary topics and political ecology themes, a complex borderland setting provides new challenges for political ecology.

Borderlands share many territorial meanings (House 1982), but can be loosely defined as the territorial regions surrounding state boundaries where livelihoods are frequently affected by the border (Newman 2006). Borderlands rely on a boundary line axis: an artificial, mutable, and humanly constructed razor's edge dividing diverse cultures, political systems, and economies (Barth 1969). The understudied Amazonian borderlands have historically contained some of the most invisible of Amazonians: in this case the borderland Asháninka.

Asháninka

The Asháninka or Ashéninka, constitute one of the largest Arawakan indigenous groups, and include at least six subgroups (Veber 2003). These people now cover a fragmented territory that once spanned approximately 100,000 km[2] in the *selva central*[1] region of the eastern slopes of the Andes (Figure 1) (Veber 2003; Benavides 2006). Historically called *Campa* in Peru and *Kampa* in Brazil, today the more modern and inclusive term of Asháninka is used to refer to all subgroups. The Asháninka are known for their ability, in the face of repeated aggression, to maintain both their ethnic identity and a high degree of elasticity in their social organization (Varese 1968). One hundred years ago, these qualities were put to the test as rubber tappers and other resource collecting *patrones*[2], or rubber bosses, captured and enslaved Asháninka before dislocating them to work in distant areas (Varese 1968). Today, Asháninka adaptability and identity continues to be tried by labor recruiters who tangle them in debt peonage. This research investigates the forgotten migration of the most invisible of the Asháninka, those borderland populations now living hundreds of kilometers from their ancestral homeland.[3]

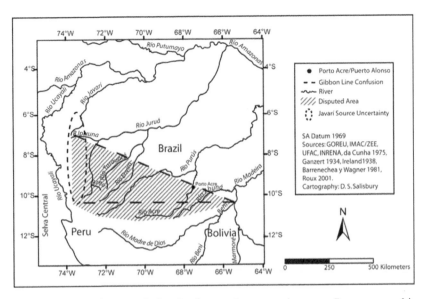

Figure 1. Boundary confusion in the southwestern Amazon. Poor geographic knowledge led to disputes between Brazil, Peru, and Bolivia.

Methods

Following Offen's (2004) historical political ecology framework, this paper relies both on library-based archival research and Amazon fieldwork. Fieldwork consisted of 10 months of research in the borderlands shared by Brazil and Peru. The specific inductive approach began with grounded local level fieldwork (Butzer 1989) investigating indigenous resource use in 2004. A research team traveled seven days upriver from the Peruvian town of Pucallpa to spend three months conducting participatory methods, ethnography, and semi-structured interviews to understand and map the resource use and resource conflicts of the borderland Asháninka. Following this, progressive contextualization (Walters and Vayda 2009) led to interviews with Asháninka relatives in Brazil and other key informants in both countries, while Global Positioning System waypoints of borderland resource management, transboundary invasions, and road building were analyzed using geographical information systems and remote sensing in both Peru and the United States.

On returning to the United States, archival research helped to clarify the historical processes underpinning the oral interviews conducted in the field. Archives provided detailed historical information on the Brazil-Bolivia borderlands, but less on the Tamaya and Juruá Rivers. However, missionary accounts, *indigenista* writings, and other archival materials helped illuminate the historical political ecology of the region. Of particular utility were the joint commission reports of the bi-national reconnaissance of the Juruá and Purús basins. The Brazilian side of the commission was led by two Brazilians of note. Euclides da Cunha, the famous writer, engineer, and proto-political ecologist (Hecht forthcoming), explored the Purús River (Comisión Mixta 1906; da Cunha 1967) before becoming one of the iconic literary figures of Brazil. General Belarmino Mendonça, largely unknown outside of Brazil, also demonstrated a keen and sensitive eye in his observations of the physical and human geography of the Juruá River (Mendonça 1907). The Peruvian counterparts for the Purús and Juruá Rivers, Captain Pedro Alexandre Buenaño and First Lieutenant Numa Pompilio León respectively, wrote less detailed descriptions, but showed equal courage in exploring the limits of the two watersheds (Comisión Mixta 1906).

Resource hunger, cartographic uncertainty, and the political boundaries of Amazonia

The 12,000 kilometers of international boundaries spanning the Amazon basin result from contests for natural resources driven by global markets. These boundaries have been adjusted and solidified with the ebb and flow of the Amazon resource booms. The first boundary adjustment, and thus dispute, led to the 1494 Treaty of Tordesillas between Spain and Portugal and the westward migration of the meridian separating the two powers (Porras and Wagner 1981; Roux 2001). Uncertain geographic knowledge of the interior, distilled into imaginative official cartographies, resulted in multiple interpretations of the Tordesillas line (Harrisse 1897), not only undermining the subsequent treaties of Madrid and Ildefonso, 1750 and 1777 respectively, but also encouraging Portuguese explorers to expand westward in search of gold, timber, spices, and other natural resources (Ireland 1938; Roux 2001).

In the 1800s, the newly independent Amazonian states agreed to largely abide by the administrative divisions, and flawed cartography of the colonial powers (Porras and Wagner 1981; Díaz Ángel 2008). The new states recognized the cartographic uncertainty, but felt the large, unpopulated, and relatively uncontested borderlands would minimize conflict (Ireland 1938). In reality, numerous "unassigned" indigenous people, missionaries and adventurers populated the borderlands, with Brazilian entrepreneurs and adventurers particularly aggressive in pushing into these border areas in search of resources, chiefly the *Hevea* rubber trees (Hemming 1978; Díaz Ángel 2008).

Land rich in *Hevea* rubber, combined with this lack of geographical knowledge, created conflict in the Southwestern Amazon (Díaz Ángel 2008). Brazil, Bolivia, and Peru tried to make sense of their boundaries while sharing a "complete early ignorance of the course and source of one river (the Javarí) and the uppermost extent of the basins of two others" (the Purús and the Juruá) (Ireland 1938, p. 130) (Figure 1). Despite this ignorance, Brazil and Bolivia signed an 1867 treaty in La Paz dissolving the Ildefonso treaty, and cutting out Peru by drawing the border between Brazil and Bolivia from the mouth of the Bení, on the parallel 10° 20″ S,[4] to the headwaters of the Javarí (Figure 1) (Ganzert 1934; Ireland 1938; Roux 2001). Instead, the Javarí's headwaters lay a full three degrees to the north of the estimate in the treaty, leading to uncertainty and conflict over the rubber-rich Juruá and Purús headwaters (Roux 2001).

Into this uncertainty poured Brazilian and foreign entrepreneurs with their indigenous, Afro-Brazilian, and European labor forces. Those indigenous people not interested in joining the grueling working conditions of this economic (and political) project sought refuge along distant tributaries and interfluvial zones upriver where the *Hevea* stands thinned out along with economic interest. The ascent of entrepreneurs up the Purús was so rapid, Bolivia established a customs house at Porto Acre to tax rubber revenue from forests upriver (Figure 1). The invading rubber tappers rejected this Bolivian claim to control and declared Acre an independent nation (Tocantins 1961). The short and idiosyncratic independence of Acre is well documented along with the bloody struggles between the Brazilian colonists and a US-backed Bolivian army (Tocantins 1961; Hecht and Cockburn 1990). Eventually, in 1903, Brazil and Bolivia signed the Treaty of Petrópolis, giving Brazil the Acre territory based on the newly fixed source of the Javarí by a Peru-Brazil bi-national commission (Ganzert 1934). Despite the Peruvian presence on the commission, the 1903 treaty ignored Peruvian claims to Acre, leading to another, less documented, conflict pertinent to the borderland Asháninka.

Biogeography as political destiny?

The physical and biological geography of the Juruá River basin helped define the political boundary separating Brazil and Peru. Between the Juruá and Ucayali rivers rose the dividing range, an isolated outcrop of mountains that became an axis of the political boundary between Brazil and Peru. This watershed divide, the *Sierra del Divisor*, was not only subtly visible on the ground, but also marked the end of territory rich in the rubber producing *Hevea brasiliensis* tree (Figure 2). Today, few *Hevea* trees exist on the Ucayali side of the divide, while rubber tappers still tap the trees of the Juruá Valley.

The range of the richest rubber producing tree, *Hevea brasiliensis,* served as a powerful dividing line. During the rubber boom, Brazilian and foreign rubber tappers expanded up the Amazon tributaries until the *Hevea brasiliensis* stands remained thick and productive, but rarely beyond. Tastevin (1920, 1926) noted the speed and limit of Brazilian ascent up the Juruá, where rubber tappers explored the entire headwaters by 1900 before retreating to the mouth of the Breu River because upriver, *Hevea* rubber stands grew sparse (Figure 2). The Alto Juruá Commission (Comisión Mixta 1906, p. 274) also noted this *Hevea* line in their

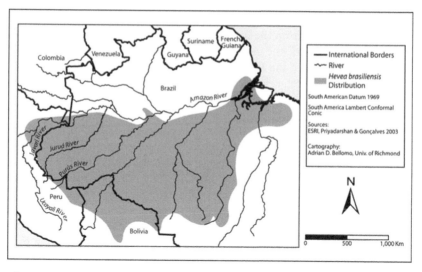

Figure 2. *Hevea* Boundaries. The relationship between current political boundaries and the distribution of *Hevea brasiliensis.*

observations of the Breu River, part of the current boundary line, "On both sides of the Breu they extract latex. From there to the headwaters the *Hevea* thins out until it disappears; leaving only *caucho* that is exploited exclusively by the Peruvians."

The blurred ecological borderlands where *Hevea* thinned and the *caucho* tree, *Castilloa ulei*, predominated, provided terrain for two distinct resource management approaches and political states to jostle. *Castilloa* rubber extraction almost always requires the felling of the tree, and thus a nomadic lifestyle and constant exploration, while *Hevea* rubber can be sustainably extracted by a sedentary rubber tapper daily walking a handful of trails to cyclically tap the same living trees (Mendonça 1907; da Cunha 1976; da Cunha 1967; Hecht 2004). While the quest for *caucho, Castilloa* rubber, brought Peruvians to these borderlands, the Peruvian government's spies recognized *Hevea* rubber as having the more lucrative and sustainable economic future (Villanueva 1902). Similarly, Brazil's da Cunha noted the unappealing nature of the harvesting methods and harvesters of *caucho,* called *caucheros,* and thus questioned the *caucheros*/Peruvian claims to territory (Hecht 2004).

Peruvian *caucheros* penetrated the *Hevea* rich territory of the *seringueiros, Hevea* rubber tappers, by taking advantage of a key feature of the borderland physical geography: the *varadero.* These portage trails led from one watershed to another: often linking headwaters. *Varaderos,* used for thousands of years by indigenous peoples before contact, became heavily trafficked by *caucheros,* spies, smugglers, and soldiers during the rubber era (Figure 3).

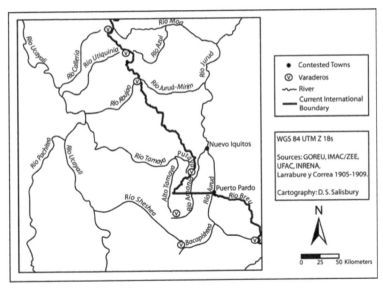

Figure 3. *Varaderos* and contested towns along the central borderlands of Peru/Brasil in the early 1900s. *Varaderos* served as crucial links between the Ucayali and Juruá basins in the early 1900s and continue to do so today.

While watershed divides and ecological gradients had a powerful impact on the political and economic geography of Amazonia, this power largely emanated from the global quest for, and preeminence placed on, natural resources. Thus, going beyond Hecht (2004), influential environmental variables could become hyper-influential in the context of global resource dynamics, for example the *Hevea* resource not only facilitated the shaping of political boundaries here, but also helped germinate cultural archetypes such as the mythologized rubber tapper analyzed by Vadjunec *et al.* (2011).

Historical conflict in the borderlands

The *varadero* connecting the headwaters of the Tamaya/Putaya and Amônia Rivers served as an important conduit in historical Brazil/Peru relations. In 1897, the first Peruvian *caucheros* coming to Brazil traveled this *varadero* (Mendonça 1907); Peruvian soldier/spy Espinar and Peruvian diplomat/spy Villanueva also used this *varadero* as well as the Juruá Mirim-Abujao *varaderos* to investigate the degree of Brazilian influence along the upper Juruá (Villanueva 1902; Espinar 1905; Rosas 1905). This *varadero* not only helped Peruvians established a base, customs office, and town named Nuevo Iquitos at the mouth of the Amônia, but also allowed Coronel Pedro Portillo to bring Peruvian soldiers to garrison this Peruvian foothold (Portillo 1905b). The Brazilian government, local authorities, and independent rubber tappers took umbrage at the Peruvian presence and tax and, in November 1904, a delegation of Brazilian soldiers and *seringueiro* volunteers attacked Nuevo Iquitos, and forced the surrender of Peruvian forces, before letting them retreat upriver to the mouth of the Breu River, Puerto Pardo, the beginning of the end of the *Hevea* stands, (Figure 3) (Mendonça 1907).

This November attack took place despite Brazil and Peru's signing of the 12 July 1904 *modus vivendi* neutralizing conflict in the upper Juruá and Purús basins and creating joint commissions to ascertain suitable boundaries. In the two years following the 1904 conflict, these commissions explored their respective headwaters with considerable help from the indigenous and rubber tapper residents (Comisión Mixta 1906; Mendonça 1907). Their detailed fieldwork helped the controversy come to a close with a signing of the treaty of boundaries, commerce, and navigation in 1909, delimiting the current boundary of Bolivia to the Javarí River. Not surprisingly, much of this boundary followed settlement patterns established according to the biogeography of *Hevea*, with global markets dictating the salience of this resource.

Hidden histories: the borderland Asháninka diaspora

This boundary became the locus for a borderland Asháninka people. During his reconnaissance of the upper Purús, da Cunha (1967) noted the indigenous majority among the "invading" Peruvians; indigenous people, particularly the Asháninka, formerly called the Campa, filled both labor and leadership positions. He observed,

> In general there are five Peruvians ... per 100 Piros, Campas, Amahuaca, Conibos, Shipibos, Samas, Coronahuas, and Yaminahuas, which one stumbles across in various states of bondage and indolence, all conquered by the shotgun, all deluded by extravagant contracts, all now yoked to the most complete slavery. (da Cunha 1976, p. 264)[5]

However, even within these contracts, da Cunha highlights the agency peculiar to the Asháninka, recognizing that the Campa were able "to preserve a primitive freedom, thanks to their ferocity, within the tortuous contracts they accept" (da Cunha 1976, p. 264). He is also impressed by the famed Asháninka leader Venancio (Santos-Granero and Barclay 2000), whom he meets on the Purús and calls "the Curaca Vinésio, or Vicenzio, and who dominates there: his influence and empire radiate over all the other headmen of the region" (da Cunha 1976, p. 199).

Da Cunha's Juruá counterpart, Colonel Mendonça, also finds agency among borderlands indigenous peoples, but of a helpful rather than domineering nature,

They were the guides and helpers of the pioneers; those that were the first to use *caucho* ... they taught the civilized medicinal and other virtues of many plants; on the Juruá they are multidimensional cultivators of both soils and fields having repeatedly given food to those who have come to usurp their lands, women and children. (Mendonça 1907, p. 141)

To Mendonça these borderland people deserved "humanitarian and sympathetic compensation" (Mendonça 1907, p. 141). Mendonça also realized their strategic importance for these remote frontiers and sought to incorporate them into his geopolitical project, "... the most efficient and perhaps least costly, to submit them to a brand of military regime, administered without severity in villages alongside the boundary lines that so suit us to garrison, without destroying their family ties ..." (Mendonça 1907, p. 142). While ultimately these borderland people slipped into obscurity after the delimitation of the border and the rubber bust, receiving neither compensation nor military employment, the Asháninka still policed the border a hundred years after Mendonça's anticipatory comments.

Archival references describe a bellicose Asháninka, perhaps the most suited border watchmen of the many indigenous groups brought by *patrones* to work these resource frontiers. Unlike some imported peoples, the Asháninka, reknowned warriors with charismatic headmen, were feared and respected in the borderlands. At the beginning of the 20th century Euclides da Cunha (1967, p. 58) recited a long list of indigenous peoples inhabiting the borderlands before ending with a description of "the warlike Campa of the Urubamba" as "above all, supplanting the rest in fame and courage ..." Others document the Asháninka's warlike ways: Franciscans suffered heavy losses to establish missions in the Asháninka homelands in the *selva central* (Varese 1968; Weiss 1975). However, resistance to the missionaries paled in comparison to the Asháninka's 1742 revolt. This uprising began with the expulsion of all missionaries, then annihilated two Spanish military companies, before maintaining over a century of resistance and independence despite repeated Spanish and Peruvian attempts to penetrate their homelands (Varese 1968; Weiss 1975). Nevertheless, the Asháninka also turned their warlike ways on themselves, raiding and enslaving their cousins to sell as rubber tappers, servants and bodyguards to Brazilians and Peruvians alike (Clark 1954; Varese 1968; Bodley 1972; Weiss 1975; ACONADIYSH 2004).

Beginning with the rubber boom and continuing today, rubber *patrones*, slavers, and others, scattered a portion of the Asháninka nation far and wide (Figure 4). Asháninka slaves and workers reached Madre de Dios, Loreto, Lima, Brazil, and Bolivia, and left their descendants (Von Hassel 1905; Clark 1954; Varese 1968; Bodley 1972; da Cunha 1976; Weiss 1975). Sometimes traded by their parents for tools (Velarde 1905), sometimes enslaved by other Asháninka (Fry 1907; Clark 1954; Varese

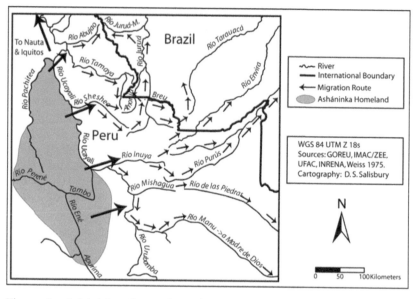

Figure 4. Asháninka dispersal to the borderlands. The Asháninka spread throughout the Peru/Brazil borderlands as laborers and security during the rubber boom.

1968; Samanez y Ocampo 1980), sometimes led by their own strong men (Portillo 1905a; da Cunha 1976; Santos-Granero and Barclay 2000), and sometimes enslaved or coerced by *patrones* (Clark 1954; Varese 1968; Bodley 1972), the current borderland Asháninka are likely remnants of this complex diaspora. However, the timing of their occupation of the borderlands remains uncertain.

This uncertainty stems not only from poor records, but also from the legendary mobility of the Asháninka (Denevan 1971, Varese 1968), which complicates defining their historical presence in any place outside their long established homelands in the *selva central*. The Brazilian Castello Branco (1950) even asserts Asháninka presence in the upper Juruá as early as the beginning of the 18th century. While this early presence remains unsubstantiated, the same source (Castello Branco 1922) more reasonably finds the Asháninka wandering the borderlands in the early 1900s.

Castello Branco's contemporary, Tastevin (1920), recognizes Asháninka presence in the eastern foothills of the *Sierra del Divisor* and along the headwaters of the Juruá Mirim River in 1920. This assertion, combined with Castello's 1922 reference, supports documentation of the rubber labor system's use of Ucayali indigenous groups in neighboring watersheds (da Cunha 1967; da Cunha 1976; Santos-Granero and Barclay 2000; Pimenta 2002; ACONADIYSH 2004). Similarly, da Cunha encountered numerous Campa along the adjacent Purús watershed in 1904–1905 (da Cunha 1967; da Cunha 1976). More recent accounts also place the arrival of the Asháninka during the rubber period (ACONADIYSH 2004) or acknowledge their presence shortly thereafter (López 1925).

Based on these sources, Figure 4 illustrates the migratory routes and *varaderos* used by the Asháninka to access the borderlands during and after the rubber boom. *Patrones* from both Brazil and Peru brought Asháninka to the borderlands, and each country blames the other for this dislocation (da Cunha 1967; da Cunha 1976; Pimenta 2002; ACONADIYSH 2004). However, others underscore Asháninka agency in driving their own migration (Santos-Granero and Barclay 2000). Following the collapse of the rubber boom in the early 20th century and the exit of the *patrones,* some Asháninka returned to their homelands in the *selva central* while others remained in the borderlands (López 1925; ACONADIYSH 2004). Therefore, this archival research demonstrates that the borderland Asháninka are a population of mixed provenance whose borderland presence and sense of place were sparked by the global thirst for rubber in the late 1800s and early 1900s.

The arrival of the current borderland Asháninka populations along Brazil's Juruá River can be traced back to the 1930s through oral interviews (Pimenta 2002). Figure 5 maps the location of the borderland Asháninka in 2004. That year, in Brazil, approximately 900 Asháninka

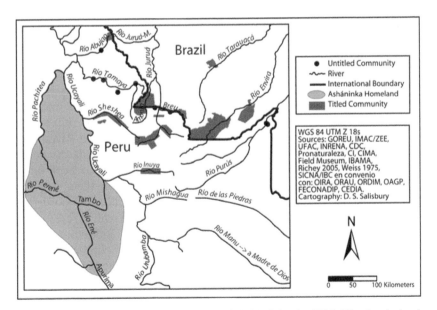

Figure 5. Distribution of the borderland Asháninka in 2004. The borderland Asháninka now reside on both sides of the border in the Peruvian and Brazilian borderlands, although many communities still remain unrecognized and untitled by the Peruvian state.

lived in five indigenous territories[6] dispersed along the effluents of the Juruá River (Pimenta 2005). The same year, across the border along the Peruvian Alto Juruá and its tributaries, 900 Asháninka lived in eight borderland communities (ACONADIYSH 2004).

The borderland Asháninka on the Tamaya and Abujao Rivers of the Peruvian borderlands trace their residence back to the 1940s. While there are no published accounts of their location or population, Richey's (2005) fieldwork found only 403 living along the Tamaya in four untitled communities and one titled community,[7] and about 50 living along the Abujao in one titled and one untitled community. In 2004, the Asháninka elders remembered their Brazilian kin, having passed through Brazil as residents either of the Juruá, Amônia, Envira or Purús Rivers (Figure 4). Figure 5 maps the Peruvian borderland Asháninka, who appear neither on the thematic maps of Tessmann (1930) nor other maps of indigenous territories in Peru (Benavides 2006).[8]

Some Asháninka populations may be remnants of the *Campa* who worked the borderlands as tappers, food producers, bodyguards, or slavers during the rubber epoch of the late 19th and early 20th century (Pimenta 2002). Others appear descended from the same, but after their forebears returned to the *selva central*, they had taken advantage of their family's expanded geographic knowledge to come to the borderlands in search of relief from slaving parties, the violence of the *Sendero Luminoso*,[9] or to find better hunting grounds. Oral interviews revealed that those Tamaya Asháninka not related to the Brazilian populations on the Amônia were either brought as slaves by *patrones* to log the forests, or had traveled with logging outfits as independent workers.

The elders among the Abujao Asháninka also tell of first working in Brazil on the Envira, Breu, and Juruá Rivers before leaving that area with the *patrón* Cristóbal Fuchs Colón, who according to one informant exclaimed one day, "we are Peruvian, we should live in Peru," and led them back to Peru and to the lower Abujao via the Sheshea River. From there they gradually migrated up the Abujao River towards the border, with some even crossing back into Brazil via *varaderos* to work on the Juruá River as rubber tappers before jumping back over the *divisor*.

Regardless of their location or country of residence, the borderland Asháninka share a complex transboundary history. Archival research and oral histories demonstrate the confluence of Hecht's (2004) three themes: by revealing the Asháninka hidden within "empty" borderland landscapes: with both history and landscapes aligned to the biogeography of *Hevea brasiliensis*. Now, the international boundary continues to be crisscrossed by the borderland Asháninka for reasons of family or commerce. However, despite their kinship and constant contact, the still hidden borderland Asháninka of Peru are very different from the borderland Asháninka of Brazil as they face the next borderland resource paroxysm: timber.

A borderland Asháninka comparison

The Peruvian Asháninka of the Alto Tamaya community and the Brazilian Asháninka of the Apiwtxa community on the Rio Amônia, two intermarried communities separated by only a score of kilometers and the Tamaya-Amônia *varadero*, have developed distinct identities and marshaled very different levels of power to face the incursion of outside loggers (Figure 6).

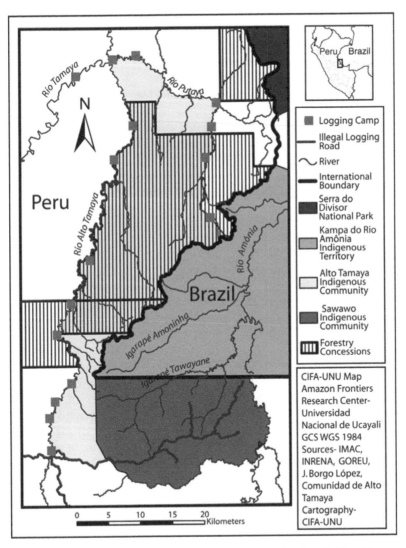

Figure 6. Logging activity in the Asháninka homelands of Alto Tamaya, Apiwtxa, and Sawawo. More details about CIFA maps from the Universidad Nacional de Ucayali are forthcoming in Salisbury 2011.

Alto Tamaya, Peru

Until 1998 the Asháninka of the Alto Tamaya and Putaya Rivers had lived in their traditionally separated and mobile family units for over fifty years. While this fragmented and mobile settlement pattern helped the indigenous group resist outsiders hundreds of years ago in the *selva central* (Varese 1968), in the modern era these isolated households created challenges for this community looking to gain titled land through the requisite demonstration of organization and occupation of territory. At the same time, not all in Alto Tamaya craved community. Some informants interviewed had migrated from Apiwtxa to take advantage of the lack of regulation regarding the harvesting of natural resources. Others came to Alto Tamaya with *patrones* looking to log these borderland forests. Nevertheless, many families became more open to organizing because of the threat of outside loggers, and the recognition of the benefits of educating their children.

In 1998, a family of undocumented and illiterate Asháninka travelled to the current village site and made a call for their kin in the watershed to join them. In forming a nucleated community, the Asháninka of Alto Tamaya, a cluster of formerly dispersed households, became more visible on the landscape and on the map, now potentially recognized by outsiders other than the local loggers, skin hunters and *patrones*. As one informant explained, "Before we lived as animals, houses over there, others over here, later we got together, organizing to educate our children." Their decision to organize was timely because, two years later, the new Peruvian Forestry law introduced forestry concessions throughout the Peruvian Amazon. This law reclassified their homelands from an "unoccupied" frontier watershed where timber extraction had been banned by the state, to a checkerboard of forestry concessions available to the highest bidder (Figure 6).

This Forestry law brought an onslaught of illegal and quasi-legal loggers waving concession permits, logging Alto Tamaya forests, and at times even employing the Asháninka to topple Tamaya trees (Figure 7). However, the greatest threat to their community was their long time *patrones*, who recognized how an Ashaninka village competed with *patrón* claims to the territory, resources, and the scarce labor of the region. As Little (2001) notes, territorial claims remain dormant until a dispute enlivens a variety of behaviors such as protest, legal appeal, violence, negotiation, and public denunciation. Thus, the community's first plea to a long time *patrón* led him to persuade several families to abandon the community. According to community leaders Rodrigo and Tanya,[10] he said:

> Don't listen to Rodrigo and Tanya; if you follow them you will end up naked in the community ... there will be no money ... everything about the community is a lie ... they will not get a teacher, they will not organize, they

Figure 7. Two borderland Asháninka children stand on lupuna (*Chorisia integrifolia*) logs harvested by others from their homeland in 2004. For policy makers, resources continue to overshadow the inhabitants of the "empty" Amazon borderlands. Photograph by David Salisbury.

will never get titled, it is all a lie, work, work the timber … here you will see money. If not, in the community, you will end up naked.

Berto, the most educated of the community, related that another *patrón* also came to the community saying, "In the community you will have nothing. Come, work with me. I will pay you. I will give you everything." But Berto no longer believed the *patrones*, concluding the conversation with, "but it is not like that, it is not like how they promise." The power of a *patrón* in the isolated Amazonian borderlands cannot be over-estimated. These powerful individuals provide money, medicine, and meals in times of duress, and some are even family, employing Asháninka women as nannies to their children, or fathering children with these or other Asháninka women.

Some Asháninka families listened to the *patrones* and walked away, while others continued to organize, travelling 5 days downriver to walk barefoot around Pucallpa delivering letters seeking recognition of their community to any organization that would listen: the Regional Director of Agriculture, the Defender of the People, the Mayor, the Admiral of Ucayali, INRENA (National Institute of Natural Resources). Living in the city and engaging the Peruvian bureaucracy was not easy for the cash poor and largely illiterate borderland Asháninka. As Tanya said, "In the city, when you don't have *plata* (cash)[11] you don't drink even a drop of water."

Progress began to be made in 2003 when the National Registry of Native Communities included Alto Tamaya, and the Ministry of Education created a community school[12] and assigned a teacher. Now a recognized community, leaders still struggled to expel the loggers. When the community convinced an INRENA forester to inspect the illegal logging in their untitled homelands, he wrote a report detailing both legal logging on the overlapping concessions, and Asháninka workers illegally logging under their *patrón's* supervision. The *patrón* was laundering this timber with the paperwork from the neighboring concessions and providing the only source of paid employment, an exploitative debt peonage. Not totally discouraged by INRENA's ambivalent report, the Asháninka wrote 15 letters between 2002 and 2004 asking to dissolve the overlapping concessions. Finally, frustrated by the lack of progress, they asked INRENA to apprehend timber en route to Pucallpa sawmills that had been logged illegally in their homeland. In 2004, INRENA seized over 275 cubic meters of hardwood, worth approximately 200,000 US dollars[13] in the storage area of a Pucallpa sawmill. However, the timber was released to the *patrón* when he coerced the illiterate vice president of the community to place his thumbprint on a document stating the timber came from elsewhere.

Finding legal channels expensive and inefficient, the Alto Tamaya Asháninka also took matters into their own hands. In 2003 delegations from Alto Tamaya blocked off three tributaries because the loggers did not give the community a percentage of the timber profits earned from the Asháninka homelands. The loggers asked the Asháninka to show title to the land, and when no documentation materialized, continued to log. The community also attempted to barricade a tributary of the Putaya River, but their chainsaw malfunctioned. The concessionaire, on hearing of their efforts, came and promised to leave a percentage of the timber extracted, but never did. In another case, the subchief of Alto Tamaya confronted a *patrona* (female *patrón*) for financially backing *tabloneros* (plank makers) who were sawing and removing mahogany planks from forests in both their homelands and Brazil. The *patrona's* response was to indicate the Asháninka had no legal title. In addition, she repeated over and over that she worked with Edgar Velásquez. The subchief did not know who Edgar Velásquez was, but in 2004, Velásquez, was a sawmill owner as well as the Governor of the entire region. Currently, the timber business continues to drive regional politics, and at the time of this writing, the Alto Tamaya Asháninka still do not have title to their lands, much less rights to the trees remaining.

The state's commodification of these trees through laws and policies simultaneously unveils a hidden social history and creates another Amazonian struggle for social justice (Hecht 2004) associated with competing territorial and resource claims (Little 2001). Once free to fish, hunt, and log in the "empty" borderlands, the untitled now must defend themselves and their resources from the state forest policy. The state's forestry system not only creates forestry concessions, and thus concessionaires, but also facilitates the international markets that empower the *patrones*. These forces wield far more power than the borderland Asháninka, in another example of asymmetrical power relations along contested Amazonian frontiers (Schmink and Wood 1992; Little 2001). Yet, across the border the struggle plays out in a very different way.

Apiwtxa, Brazil

The incursion of loggers into the Peruvian borderland territory of Alto Tamaya mirrored a similar invasion of the neighboring Asháninka territory in Brazil now called Apiwtxa. There, Brazilian loggers began arriving in 1970, and their numbers peaked during the late 1980s Brazilian timber boom. During this resource scramble, Brazilian *patrones* enmeshed the Apiwtxa Asháninka in a debt peonage system that tightened as the demand for high-grade hardwood increased and supply diminished. Formerly invisible and dispersed in the Brazilian borderlands, the Asháninka organized to resist Brazilian loggers who used tractors to extract more than 2,500 logs (CEDI 1993) across 80 kilometers of newly created logging roads (Pimenta 2002). In 1992, after much struggle, the Asháninka of Apiwtxa obtained their title with the help of anthropologists and indigenistas (Pimenta 2002). Seven years later, the Asháninka of Apiwtxa would organize again, but this time with a title in hand, and the power of the Brazilian flag waving behind them.

The residents of Apiwtxa began anticipating trouble on their southern border in 1999 when their Peruvian cousins in Sawawo agreed to a logging company proposal to establish a road from the Ucayali River, and to partner in the selective logging of their territory (Figure 6) (Pimenta 2002; Hazera and Salisbury 2008). In December of 2000, the Asháninka of Apiwtxa alerted Brazilian authorities of trespassers. According to the press, in an attempt to avoid bloodshed, army helicopters beat an Asháninka war party to the logging site to discover eight clearings in Brazilian territory along with a network of logging trails leading from Peru (Pimenta 2002). This helicopter dash echoes the commissions of Mendonça and da Cunha of one hundred years earlier who also "sprinted" to the border in an effort to reconnoiter resource wars and mitigate conflict.

A month later, one of the Apiwtxa leaders was quoted in the *Folha do São Paulo* (2001, p. A2), Brazil's most widely circulated newspaper, as saying, "We want everything resolved peacefully but if nature is at risk and nothing is done, we will kill and die fighting for our people." While the attention from Brazilian authorities slowed trespassing, the logging company extracted over 6,500 m^3 of mahogany and tropical cedar in their first two harvests of Sawawo territory (Aquino, T. 2004).[14] By 2002, the logging company's road from the Ucayali River extended over 140 kilometers (Hazera and Salisbury 2008). This road continues to advance parallel to the border, and according to the President of Apiwtxa, will continue to target borderland indigenous territories on both sides of the Juruá River (Hazera and Salisbury 2008).

Despite the aforementioned attention of the Brazilian state, the invasions persisted. In November of 2002 Apiwtxa discovered and burned a Peruvian camp a kilometer inside their land (Martins 2003). A year later, in November of 2003, the Apiwtxa Asháninka found 60 Peruvians logging mahogany and tropical cedar with tractors and chainsaws inside their territory (Schneider 2003). In January of 2004, the Asháninka convinced a team of army, federal police, the Federal Ministry of the Environment (IBAMA) and the State Ministry of Environment (IMAC) to follow up this discovery with aerial and terrestrial reconnaissance. The reconnaissance documented logging roads shadowing the Brazilian border along with felled mahogany and ten-meter-wide skid trails inside Brazilian territory (Martins and Freddo 2004). A month later, the Brazilian army and indigenous agency (FUNAI) forced 30 Peruvian loggers out of Apiwtxa land (FUNAI 2004). Nevertheless, trespassing continued from both the south and west, and in March of 2004 a federal judge of Acre held the federal police, IBAMA, and FUNAI responsible for not complying with promises made to the Asháninka. The judge also ordered the three institutions to update the border monuments and set up outposts in the region (Instituto Socio Ambiental 2004b). Four months later, the Military Command of Amazonia announced the forthcoming establishment of three new military bases in the borderlands with one planned just outside of Apiwtxa (Figure 6) (Maia 2004).

The establishment of the promised infrastructure proceeded with the delays typical of the border region until September 2004 when the University of Brasília hosted a four-day celebration of Asháninka culture in the Brazilian capital. This celebration gave the Asháninka a stage to call attention to the continued trespassing into Brazilian and Asháninka territory (Amorim 2004). While in Brasília, the Asháninka personally denounced the Peruvian invasions to Marina Silva, the Brazilian Minister of Environment,[15] FUNAI, a federal judge, a pop music star, a representative of the Protection System of Amazonia (SIPAM), and members of the Brazilian press (Instituto Socio Ambiental 2004a). The charismatic and eloquent leaders of Apiwtxa made an immediate impression with Marina Silva saying, "In terms of organization, the Ashaninka are an example for all of Brazil," (Ministério do Meio Ambiente 2004).

That the Apiwtxa Asháninka impressed the then Minister of the Environment, and fellow Acre native, Marina Silva came as no surprise. Since titling their land the Asháninka of Apiwtxa have: (1) created a cooperative for sustainable harvesting of natural resources, (2) developed a school for traditional ecological knowledge, (3) won prizes for community-based conservation, and (4) starred in, filmed, and directed videos about their life in harmony with the forest (Apiwtxa 2010). Many of these initiatives have flourished through Apiwtxa's close and lengthy relationship with nongovernmental organizations such as the Commissão Pro-Índio (CPI), but the charisma, imagination, and initiative of Apiwtxa community members, and particularly their leadership, appears extraordinary. Given a stage in Brasília, the heart of Brazil, the Apiwtxa leadership's charismatic presence and persuasive message of a national security threat galvanized Brazil's security and environmental forces into immediate action.

On 27 September 2004 a Brazilian joint operation apprehended four Peruvian loggers, burned a logging camp, and dynamited captured mahogany (Sales, V. 2004). On 12 October 2004 the same operation seized Peruvian loggers in Asháninka territory and just to the north, inside the Serra do Divisor National Park, 150 m^3 of mahogany (Campos 2004). Three days later the operation captured 700 m^3 of tropical cedar Peruvian loggers had abandoned in the Serra do Divisor National Park (Sales, A. 2004). That weekend, government authorities met with the Asháninka in Apiwtxa and presented them with a GPS and satellite phone to alert them if the invasions continued (Sales, A. 2004). In October, Operation PeBra (Perú Brasil) resumed, capturing 26 loggers in the Serra do Divisor National Park (Antunes 2004; O Rio Branco 2004). This detailed history of invasions and intervention demonstrates the lengths and methods needed to capture, recapture, and focus the attention of the Brazilian state on a distant borderland. Even endowed with extraordinary leadership, influential partners, and the discourse of a national security threat, the Asháninka required all their resources and more to bring the authorities to bear.

However, the newly active border policing also had negative impacts for the community and individuals of Apiwtxa. The President of Apiwtxa twice received death threats from Brazilians in league with the Peruvian loggers (Piedrafita 2004). Other threats also led the Asháninka and Brazilian delegation to cancel their participation in a University of Ucayali-led meeting to address the invasions. At the same time, Asháninka recognition by Acre and Brazilian institutions improved markedly with the President of Apiwtxa receiving the Brazilian human rights award in 2004 (Piedrafita 2004), and the unprecedented visit of the head of the Military Command of Amazonia and the Governor of Acre to Apiwtxa in February 2005[16] to promise continued support (Maia 2005).

In March 2005, the Governor of Acre took the President of Apiwtxa with him to a meeting with President Toledo[17] in Lima, thus forcing the Peruvian head of state to recognize the borderland indigenous people affected by the planned increase in commerce and infrastructure between the two nations (*La República* 2005). Months later, continued trespassing led to the July 2005 initiation of the promised Timbó III military operation putting 7,000 soldiers in simultaneous action across three borderland states (Simonetti 2005). This Asháninka-sparked operation netted 700 m^3 of logs and 40 Peruvians involved in the illegal transport of timber (Simonetti 2005).

The success and visibility of the Apiwtxa case presents a stark contrast to the unseen struggles of the Tamaya Asháninka. The Brazilian media's highlight reel of international trespassing, capture, and indigenous defenders of Brazil overshadows the more methodical exploitation by loggers on the Peruvian side. The asymmetrical comparison supports Le Billon's (2008) argument that mainstream interpretations of resource wars focus on the sensational rather than the slower more systematic form of exploitation experienced in Alto Tamaya. The Apiwtxa Asháninka mobilize the military, meet heads of state and ministers, and successfully, if repeatedly, defend their territory whereas the Tamaya Asháninka struggle for recognition a few kilometers away. Why are the Apiwtxa so much more successful than their cousins across the border? To this there are multiple answers: higher levels of education, more effective and charismatic leadership, excellent contacts with the government and NGOs, the ability to articulate environmentalist and national security discourses, residency in a more "progressive" state and a longer history as an organized community. However, the foundation of their success lies in their formal land title, the title that gives them the authority to command the attention of the state, and the motivation to defend their territory, and thus Brazil's.

Conclusion

This brief history of the struggle along the Ucayali and Juruá divide outlines the place of international boundaries within the historical and current political ecology of Amazonia. Hecht (forthcoming) calls the scramble for territory between the rubber tappers of Brazil and the *caucho* harvesters of Peru a competition between political ecologies. This and other resource wars helped forge and then temper the international boundaries of Amazonia, giving both historical context and current applications to a borderland research approach called transboundary political ecology. This approach includes modern political geography's nuanced understanding of political borders as "part of an ongoing dynamic process" (Newman 2006, p. 156) where the multiple meanings of these not-so-static lines are understood over time, space, and scale. As a dynamic process, these boundaries provide people and institutions with both challenges and opportunities, influencing resource management decisions at a variety of scales. Furthermore, the cultural and political blurring and blending around these boundaries (Augelli 1980; Minghi

1991) create borderscapes where networks of family, friendship, and entrepreneurial connections help residents negotiate and renegotiate livelihoods (Baud 2000). To date, these borderland landscapes, livelihoods, and identities have been overlooked, despite Amazonia's extended integration into global processes. Continued close attention to the ties between cultural geography and cultural and political ecology will insure these people do not again disappear into the margins of Amazonia.

This Ucayali-Juruá longitudinal case study uses this transboundary political ecology approach to understand how the global hunger and regional quest for resources creates the political boundary, forms a transboundary people, the borderland Asháninka, and dictates the political economies the borderland Asháninka must negotiate to protect their people, homelands, and livelihoods. The research triangulates the three themes of Hecht's (2004) political ecology of Amazonia (the mythic Amazon, scientific Amazon, and the production of landscape) to find an extraordinary degree of overlap between the three in the borderlands: the plunder of natural resources in "empty" landscapes; the binding of biogeographical and political boundaries in the borderlands; and the emergence of the previously hidden Peruvian Asháninka from borderland landscapes.

Preeminent in this relationship is the global demand for Amazonian resources. Waves of resource booms drove explorers, tappers, loggers and others into the "empty" forest, first pushing proposed Portuguese boundaries west into Spanish territory. Then, the soaring price of rubber forced formerly uncertain boundaries between the Juruá and Ucayali basins to harden along ecological/economic gradients while simultaneously coercing/enticing the Asháninka peoples to migrate to the borderlands. While political boundaries attempted to follow ecological gradients, economic and political structures dictated resource possibilities, rather than a biogeographical determinism, and made those gradients of elevated importance for human use. However, the subsequent fall of the world price for rubber, forced these boundaries to become little more than an afterthought, leaving the borderland peoples forgotten in the forest. Roughly a century later, the timber boom, and the new forestry law, incited a new wave of resource collectors to surge into the "empty" forest. There, the forgotten borderland peoples mobilized to defend their homelands.

If globally driven resource plundering is preeminent, the agency and interaction of the physical geography and local people is also of critical importance to the formation and defense of the boundary lands. Within Hecht's (2004) scientific Amazon, the ecological gradient of the *Hevea brasiliensis* tree imposed limits on where rubber could be extracted and ultimately, in this case, influenced where the boundary line should be drawn. Similarly, the line could neither be drawn nor defended without the guidance of local people. Mendonça (1907) foreshadowed the role of the indigenous borderlanders a century ago when he realized the importance of local borderland knowledge and the utility of an indigenous border patrol.

Nevertheless, the transboundary comparison of the Asháninka people underscores the importance of legal land title in the face of invaders. While many other factors are in play, official land title has allowed the Apiwtxa indigenous people to jump scale from "obstacle to development" to "defender of the state" in 15 years. Apiwtxa sounded the alarm to mobilize thousands of troops on the border, even as Alto Tamaya struggled to obtain their first community school teacher. As development and resource frontiers advance and threaten, the borderland Asháninka of Peru must get their land title if their hidden social history is to ever become more than just history. Similarly, land title may be necessary for the Peruvian Asháninka to mitigate the transboundary impacts of Peruvian forestry policy, thus protecting their cousins and the forest as a populated buffer zone (Nepstad *et al.* 2006).

While the detailed fieldwork and archival research shared in this article serves as a platform for the Asháninka of Alto Tamaya to argue for title, the analysis also informs borderland research along poorly understood frontiers. Hecht's (2004) Amazonian political ecology themes can be more broadly understood as frontier silences (Harley 1988), scientific narratives, and hidden frontier histories. As such, a transboundary political ecology approach can triangulate these themes along other borders and inform conservation and environmental/social justice today by anchoring current borderland processes in an understanding of the past (Offen 2004).

Acknowledgements

This research was supported by funding from Fulbright-Hays, The Nature Conservancy, ProNaturaleza, the University of Richmond, and the Universidad Nacional de Ucayali. We would like to thank Greg Knapp, Karl Butzer, Bill Doolittle, Ken Young, Peter Dana, and Susanna Hecht for their comments on previous drafts, and Editor-in-Chief Alyson Greiner and two anonymous reviewers of the *Journal of Cultural Geography* for their insights and corrections. Particular thanks to Jackie Vadjunec and Marianne Schmink for their organization and enthusiasm. Also, thanks to our colleagues in both Brazil and Peru who made our fieldwork possible and enriched the analysis within. Finally, above all, we are in debt to the people of Alto Tamaya. We hope this small contribution might help them in their struggle for recognition.

Notes

1. Santos-Granero and Barclay (1998, p. 1) define the *selva central* as the "central jungle," "central *montaña*," or central portion of the Peruvian *selva alta* or high jungle along the eastern slopes of the Andes.

2. The *patrones* are the mixed-blood rural elite who control rural labor with financial backing, personal magnetism, and by limiting alternative options for laborers.

3. We use the simple yet powerful definition provided by Nostrand and Estaville (1993, p. 4), "A land that a group of people love to the degree that they call it a home." We first use homeland to refer to the *selva central* the Asháninka have lived in and fought for over centuries (see Figures 3 and 4), but we later use the term for the borderlands the Asháninka documented in this paper have occupied over the last 80 or more years.

4. This line was called the Gibbon Line because it took advantage of the US reconnaissance of Amazonia and US Lieutenant Gibbon's coordinates for the mouth of the Bení River.

5. All translations from the Spanish and Portuguese, unless otherwise noted, are made by Salisbury. This one, however, is adapted from Hecht (2004).

6. Terra Indígena Kampa e isolados do rio Envira (262 Asháninka residents), Terra Indígena Kampa do igarapé Primavera (21 Asháninka), Kampa do rio Amônia (472 Asháninka), and Terra Indígena Kaxinawá/Ashaninka do rio Breu (114 Asháninka) (Pimenta 2005). In 2002 there were 52 Asháninka living in the Terra Indígena Jaminawá do rio Envira (dos Santos de Almeida 2002).

7. Emory Richey's census of the 4 untitled Asháninka villages in the Tamaya watershed in 2005 showed the following numbers: San Miguel de Chambira (83 residents in 2005), Alto Tamaya (82), Nueva California (55), Nueva Amazonia de Tomajao (57), and Cametsari Quipatsi (126).

8. Exceptions are the titled communities of San Mateo on the Alto Abujao River and San Miguel de Chambira on the Tamaya River. The Instituto del Bien Común has begun mapping the borderland Asháninka of Perú while the Government of Acre has maps of the Brazilian Asháninka's titled territories.

9. The *Sendero Luminoso,* or Shining Path, was a revolutionary Marxist group fighting in the countryside in the 1980s and 1990s.

10. Pseudonyms are used throughout to protect identities.

11. "Plata," literally "silver," is slang for cash.

12. On hearing of their desire for a school, a patron asked, "Why do you want a school? To teach your children how to be thieves?"

13. This sum is based on domestic market prices on www.globalwood.com for November of 2004. Timber is priced as machine-dried at international specifications for length and quality.

14. At 2004 prices and depending on quality, these harvests of mahogany and cedar could be worth over 7 million dollars in Lima.

15. Marina Silva is a native of Acre and the daughter of rubber tappers.

16. *Página 20's* newspaper article about this landmark event contains a wonderfully illustrative photo of army personnel, Governor Viana and Asháninka leaders talking in the Asháninka village of Apiwtxa. Available online at http://www2.uol.com.br/pagina20/22022005/especial.htm.

17. A fascinating photograph of the President of Peru shaking hands with the President of Apiwtxa while the Governor of Acre, Brazil looks on is available for viewing through *La República* (*La Republica* 2005) at http://www.larepublica.com.pe/component/option,com_contentant/task,view/id,70294/Itemid,0/.

References

ACONADIYSH (Asociación de comunidades nativas para el desarrollo integral de Yurua Yono Sharakoiai), 2004. *Plan de vida de los pueblos indígenas de Yurua, 2004–2009*. Pucallpa, Ucayali, Peru: ACONADIYSH.

Aldrich, S.P., *et al.*, 2006. Land-cover and land-use change in the Brazilian Amazon: Smallholders, ranchers, and frontier stratification. *Economic Geography*, 82, 265–288.

Amorim, D., 21 September 2004. Dentro dos costumes indígenas. *Web Page Universidade De Brasília*. Available from: http://www.unb.br/acs/unbagencia/ag0904-45.htm [Accessed 13 March 2007].

Antunes, F., 2004. Peruanos são presos tirando madeira nobre em Thaumaturgo. *A Tribuna*. 23 October. Available from: http://www.jornalatribuna.com.br/w827.htm [Accessed 13 March 2007].

Apiwtxa: associação Ashaninka do Rio Amônia, 2010. Apiwtxa blog. Available from: http://apiwtxa.blogspot.com/ [Accessed April 10 2010].

Aquino, T., 2004. No tempo das invasões Peruanas. *Pagina 20*, 21 November, sec. Papo de Indio, col. 1. Available from: http://www2.uol.com.br/pagina20/ [Accessed 13 March 2007].

Augelli, J.P., 1980. Nationalization of Dominican borderlands. *Geographical Review*, 70 (1), 19–35.

Barth, F., 1969. Introduction. *In*: F. Barth, ed. *Ethnic groups and boundaries: the social organization of culture difference*. London: George Allen & Unwin, 9–38.

Baud, M., 2000. State building and borderlands in Latin America. *In*: P. van Dijck, A. Ouweneel, and A. Zoomers, eds. *Fronteras: towards a borderless Latin America*. Amsterdam: CEDLA, 41–82.

Benavides, M., 2006. *Atlas de comunidades nativas de la Selva Central*. Lima: Instituto del Bien Común.

Bodley, J., 1972. *Tribal survival in the Amazon: the Campa case*. IWGIA Document, 5. Copenhagen: IWGIA.

Brown, J. and Purcell, M., 2005. There's nothing inherent about scale: political ecology, the local trap, and the politics of development in the Brazilian Amazon. *Geoforum*, 36, 607–624.

Butzer, K., 1989. Cultural Ecology. *In*: G. Gaile and C. Willmott, eds. *Geography in America*. Columbus: Merrill, 192–208.

Campos, T., 2004. Sete madeireiros peruanos são presos em flagrante na Serra do Divisor. *A Gazeta*. 14 October. Available from: http://www.agazeta-acre.com.br/Web/Principal.jsp [Accessed 13 March 2007].

Castello Branco, J., 1922. O Juruá Federal. *Congreso Internacional De Historia De América: Annaes Do Congresso Internacional De Historia Da America*, 9, 587–722.

Castello Branco, J., 1950. O gentio Acreano. *Revista Do Instituto Histórico e Geográfico Brasileiro*, 207, 3–78.

Centro Ecuménico de Documentação e Informação (CEDI), 1993. *"Green gold" on Indian land: logging company activities on indigenous land in the Brazilian Amazon*. São Paulo, Brazil: CEDI.

Chapman, M.D., 1989. The political ecology of fisheries depletion in Amazonia. *Environmental Conservation*, 16, 331.

Clark, L., 1954. *The rivers ran east*. London: Hutchinson.

Comisión Mixta Peruano-Brasilera de Reconocimiento del Alto Purús y Alto Yuruá, 1906. *Informes de las comisiones mixtas Peruano-Brasileras encargadas del reconocimiento de los Ríos Alto Purús i Alto Yuruá de conformidad con el acuerdo provisional de Rio Janeiro de 12 de julio de 1904.* Lima: Oficina Tip. de "La Opinion nacional,".

da Cunha, E., 1967. *Á margem da história.* São Paulo: Editôra Lello Brasileira.

da Cunha, E., 1976. *Um paraíso perdido: reunião dos ensaios amazônicos.* Petrópolis: Editora Vozes.

Denevan, W.M., 1971. Campa subsistence in the Gran Pajonal, eastern Peru. *Geographical Review*, 61, 496–518.

Díaz Ángel, S., 2008. *Contribuciones a la historia de la cartografía en Colombia: Una red de investigadores y un caso de estudio.* Thesis (MA) Universidad Nacional de Colombia-Bogotá. Available from: http://razoncartografica. wordpress.com/articulos-web/ [Accessed 13 July 2010].

dos Santos de Almeida, Líbia Luiza, 2002. Asháninka: Povo austero e festivo. *Amazon Link* Web page. Available from: http://www.amazonlink.org/amazonia/culturas_indigenas/povos/ashaninka.html [Accessed 11 August 2006].

Espinar, F., 1905. Comisión científica al Río Yuruá: informe del capitán de Navío Espinar. *In*: C. Larrabure y Correa, ed. *Colección de leyes, decretos, resoluciones i otros documentos oficiales referentes al departamento de Loreto [1777–1908].* Vol. 3. Lima: Imp. de "La Opinión nacional," 410–418.

Folha de São Paulo, 2001. Frases Ashaninka. *Folha De São Paulo*, 3 January, sec. Opinião, p. A2. Available from: http://www.folha.uol.com.br/ [Accessed 13 March 2007].

Fry, C., 1907. Diario de los viajes i exploracion los ríos Urubamba, Ucayali, Amazonas, Pachitea i Palcazu por don Carlos Fry. *In*: C. Larrabure y Correa, ed. *Colección de leyes, decretos, resoluciones i otros documentos oficiales referentes al departamento de Loreto [1777–1908].* Vol. 11. Lima: Imp. de "La Opinión nacional," 369–586.

FUNAI, 2004. Funai, exército e PF expulsam madeireiros peruanos do Acre. 10 March. Available from: http://www.funai.gov.br/ultimas/noticias/1_semestre_2004/Mar/un0310_001.htm [Accessed 13 March 2007].

Ganzert, F., 1934. The boundary controversy in the upper Amazon between Brazil, Bolivia, and Peru, 1903–1909. *The Hispanic American Historical Review*, 14 (4), 427–49.

Harley, J., 1988. Silences and secrecy: the hidden agenda of cartography in early modern Europe. *Imago Mundi*, 40, 57–76.

Harrisse, H., 1897. *The diplomatic history of America its first chapter 1452–1493–1494.* London: B.F. Stevens.

Hazera, F. and Salisbury, D., 2008. *A qualitative analysis of illegal road expansion in the central borderlands of Peru.* Poster presented at: Association of American Geographers Conference, Boston, MA.

Hecht, S., 1985. Environment, development and politics: capital accumulation and the livestock sector in Eastern Amazonia. *World Development*, 13, 663–684.

Hecht, S., 2004. The last unfinished page of Genesis: Euclides da Cunha and the Amazon. *Historical Geography*, 32, 43–69.

Hecht, S., Forthcoming. *Euclides da Cunha and the scramble for the Amazon.* Chicago: University of Chicago Press.

Hecht, S. and Cockburn, A., 1990. *The fate of the forest developers, destroyers and defenders of the Amazon*. New York: Harper Perennial.

Hemming, J., 1978. *Red gold the conquest of the Brazilian Indians*. Cambridge: Harvard University Press.

House, J., 1982. *Frontier on the Rio Grande: A political geography of development and social deprivation*. New York: Clarendon Press.

Instituto Socio Ambiental, 2004a. A arte, a sabedoria, o pioneirismo e as dificuldades vividas pelos ashaninka. *Noticias SocioAmbientais*. 17 September. Available from: http://www.socioambiental.org/nsa/detalhe?id=1824 [Accessed 13 March 2007].

Instituto Socio Ambiental, 2004b. Índios Ashaninka querem garantir seu território e a segurança de suas famílias. *Instituto Socio Ambiental*. 6 October. Available from: http://www.socioambiental.org/nsa/detalhe?id=1831 [Accessed 13 March 2007].

Ireland, G., 1938. *Boundaries, possessions, and conflicts in South America*. Cambridge: Harvard University Press.

La República, 2005. Interoceánica estará lista a más tardar en junio: Presidente Toledo recibe en Palacio de Gobierno a gobernador del Estado brasileno de Acre. *La República*. 8 March. Avaliable from: http://www.larepublica.com.pe/component/option,com_contentant/task,view/id,70294/Itemid,0/ [Accessed 13 March 2007].

Larrabure y Correa, C., ed. *Colección de leyes, decretos, resoluciones i otros documentos oficiales referentes al departamento de Loreto [1905–1909]*. Lima: Imp. de "La Opinión nacional."

Le Billon, P., 2008. Diamond wars? Conflict diamonds and geographies of resource wars. *Annals of the Association of American Geographers*, 98 (2), 345–372.

Little, P., 2001. *Amazonia: territorial struggles on perennial frontiers*. Baltimore: Johns Hopkins University Press.

López, R., 1925. *En la frontera oriental del Perú*. Belem do Pará, Brasil: Unpublished book.

Maia, T., 2004. Exército começa a se instalar em áreas isoladas da fronteira. *Página 20*, 29 July, sec. Politica, col. 1. Available from: http://www2.uol.com.br/pagina20/ [Accessed 13 March 2007].

Maia, T., 2005. A presença definitiva do estado Brasileiro na fronteira. *Página 20*, 22 February, sec. Enviado Especial. Available from: http://www2.uol.com.br/pagina20/22022005/especial.htm [Accessed 13 March 2007].

Martins, M., 2003. *Relatório IBAMA Invasao da Terra Indigena Ashaninka*. Cruzeiro do Sul, Acre, Brasil: Unpublished IBAMA report.

Martins, M. and Freddo, A., 2004. *Denúncia de desmate ilegal em terra indígena Ashaninka por parte de Peruanos*. Cruzeiro do Sul, Acre, Brazil: IBAMA, IMAC Unpublished IBAMA report.

Mendonça, B., 1907. Relatorio do Commissario Brasileiro. *In*: Commissão Mixta Brasileiro-Peruana, and Belarmino Mendonça, eds. *Memoria da Commissão Mixta Brasileiro-Peruana de reconhecimento do Rio Juruá e relatorio do Commissario Brasileiro 1904–1906*. Rio de Janeiro: Imprensa Nacional, 3–212.

Minghi, J.V., 1991. From conflict to harmony in border landscapes. *In*: D. Rumley and J.V. Minghi, eds. *Geography of border landscapes*. New York: Routledge, 1–14.

Ministério do Meio Ambiente, 2004. Madeireiros peruanos invadem o Brasil atrás de madeiras nobres. 20 September. Available from: http://www.ecoagencia. com.br/a2/_a2/000001a9.htm [Accessed 14 March 2007].

Nepstad, D., et al., 2006. Inhibition of Amazon deforestation and fire by parks and Indigenous lands. Conservation Biology, 20, 65–73.

Neumann, R.P., 2005. Making political ecology. London: Hodder Arnold.

Neumann, R.P., 2009. Political ecology: theorizing scale. Progress in Human Geography, 33, 398–406.

Newman, D., 2006. The lines that continue to separate us: borders in our "borderless" world. Progress in Human Geography, 30, 143–161.

Newman, D. and Paasi, A., 1998. Fences and neighbours in the postmodern world: boundary narratives in political geography. Progress in Human Geography, 22, 186–207.

Nostrand, R. and Estaville Jr., L., 1993. Introduction: the homeland concept. Journal of Cultural Geography, 13 (2), 1–4.

Offen, K., 2004. Historical political ecology: an introduction. Historical Geography, 32, 19–42.

O Rio Branco, 2004. Operação Pebra prende mais 26 peruanos no Juruá. O Rio Branco. 24 October. Available from: http://www.oriobranco.com.br/ [Accessed 13 March 2007].

Peet, R. and Watts, M., eds. 2004. Liberation ecologies: environment, development, social movements. London: Routledge.

Piedrafita Iglesias, M., 2004. Direitos Humanos: e os dos Ashaninka? Página 20, 23 December, sec. Via Publica, col. 1. Available from: http://www2.uol.com.br/ pagina20/ [Accessed 13 March 2007].

Pimenta, J., 2002. "Índio não é todo igual": A construção Ashaninka da história da política interétnica. Thesis (Ph.D.) Universidade de Brasília.

Pimenta, J., 2005. "Instituto Socioambiental, Povos Indígenas do Brasil Ashá- ninka." Web page. Available from: http://www.socioambiental.org/pib/epi/ ashaninka/loc.shtm [Accessed 13 March 2007].

Porras Barrenechea, R. and Wagner de Reyna, A., 1981. Historia de los límites del Perú. Lima: Editorial Universitaria.

Porro, R., 2005. Palms, pastures, and swidden fields: the grounded political ecology of "agro-extractive/shifting-cultivator peasants" in Maranhão, Brazil. Human Ecology, 33 (1), 17–56.

Portillo, P., 1905a. Exploración de los ríos Apurímac, Ene, Tambo, Ucayali, Pachitea i Pichis por el prefecto de Ayacucho, coronel Pedro Portillo- Diario de la Expedición- 31 de Diciembre 1900. In: C. Larrabure y Correa, ed. Colección de leyes, decretos, resoluciones i otros documentos oficiales referentes al departamento de Loreto [1777–1908]. Vol. 3. Lima: Imp. de "La Opinión nacional,", 463–550.

Portillo, P., 1905b. Viaje del prefecto de Loreto, Coronel Pedro Portillo, al río Yuruá-20 Mayo de 1903. In: C. Larrabure y Correa, ed. Colección de leyes, decretos, resoluciones i otros documentos oficiales referentes al departamento de Loreto [1777–1908]. Vol. 4. Lima: Imp. de "La Opinión nacional,", 224–225.

Priyadarshan, P. and Goncalves, D., 2003. Hevea gene pool for breeding. Genetic Resources and Crop Evolution, 50, 101–114.

Richey, E., 2005. 47 days on the Tamaya River: October–November, 2005. Austin, TX: unpublished manuscript.

Robbins, P., 2003. Political ecology in political geography. *Political Geography*, 22, 641–645.

Robbins, P., 2004. *Political ecology: a critical introduction*. Malden, MA: Blackwell.

Rosas, B., 1905. Exploración del Yuruá-Mirim por don Ignacio Espinar Meza i Don Belisario Rosas: Diario del viaje llevado por Don Belisario Rosas- Enero de 1898. *In*: C. Larrabure y Correa, ed. *Colección de leyes, decretos, resoluciones i otros documentos oficiales referentes al departamento de Loreto [1777–1908]*. Vol. 3. Lima: Imp. de "La Opinión nacional,", 426–430.

Roux, J., 2001. From the limits of the rrontier: or the misunderstandings of Amazonian geo-politics. *Revista De Indias*, 61 (223), 513–39.

Sales, A., 2004. Exército e Ibama apreendem 700 m3 de madeira no PNSD. *O Rio Branco*. 24 October. Available from: http://www.oriobranco.com.br/ [Accessed 13 March 2007].

Sales, V., 2004. Polícia Federal prende madeireiros peruanos na reserva dos Ashaninka. *Página 20*, 28 September, sec. Especial, col. 1. Available from http://www2.uol.com.br/pagina20/ [Accessed 13 March 2007].

Salisbury, D., Forthcoming 2011. GIS maps and the Amazon borderlands. *In*: K. Offen and J. Dym, eds. *Mapping Latin America*. Chicago: University of Chicago Press.

Samanez y Ocampo, J.B., 1980 [1885]. *Exploración de los ríos peruanos Apurimac, Eni, Tambo, Ucayali y Urubamba, hecha por José B. Samanez y Ocampo, en 1883 y 1884 diario de la expedición y anexos*. Lima, Perú: Consuelo Sámanez Ocampo e Sámanez e hijas.

Santos-Granero, F. and Barclay, F., 1998. *Selva central history, economy, and land use in Peruvian Amazonia*. Washington, DC: Smithsonian Institution Press.

Santos-Granero, F. and Barclay, F., 2000. *Tamed frontiers: economy, society, and civil rights in upper Amazonia*. Boulder, CO: Westview Press.

Schmink, M. and Wood, C., 1987. The "political ecology" of Amazonia. *In*: P. Little and M. Horowitz, eds. *Lands at risk in the third world: local-level perspectives*. Boulder: Westview Press, 38–57.

Schmink, M. and Wood, C., 1992. *Contested frontiers in Amazonia*. New York: Columbia University Press.

Schneider, F., 2003. Peruanos invadem área indígena no Acre. *Página 20*, 21 November, sec. Enviado Especial. Available from: http://www2.uol.com.br/ pagina20/ [Accessed 13 March 2007].

Simmons, C.S., *et al.*, 2007. Amazon land war in the south of Pará. *Annals of the Association of American Geographers*, 97 (3), 567–592.

Simonetti, A., 2005. A Amazônia sob vigilância. *Amazonas Em tempo*. 21 July, Available from: http://www.emtempo.com.br/ [Accessed 13 March 2007].

Tastevin, C., 1920. Le fleuve Juruá (Amazonie). *La Géographie*, 33, 11–34.

Tastevin, C., 1926. Le haut Tarauacá. *La Géographie*, 45, 34–54.

Tessmann, G., 1930. *Die Indianer nordost-Perus grundlegende forschungen für eine sytematische kulturkunde*. Veröffentlichung Der Harvey-Basslerstiftung. Hamburg: Friedrichsen, de Gruyter.

Tocantins, L., 1961. *Formação histórica do Acre*. Rio de Janeiro: Conquista.

Vadjunec, J. and Rocheleau, D., 2009. Beyond forest cover: land use and biodiversity in rubber trail forests of the Chico Mendes Extractive Reserve. *Ecology and Society*, 14 (2), 29.

Vadjunec, J., Schmink, M. and Gomes, C.V.A., Forthcoming 2011. Rubber tappper citizens: emerging places, policies, and shifting identities in Acre, Brazil. *Journal of Cultural Geography*.

Varese, S., 1968. *La sal de los cerros: notas etnográficas e históricas sobre los campa de la selva del Perú.* Universidad Peruana De Ciencias y Tecnología. Departamento De Publicaciones. Antropología. Lima: Universidad Peruana de Ciencias y Tecnología.

Veber, H., 2003. Ashaninka messianism: the production of a "black hole" in Western Amazonian ethnography. *Current Anthropology*, 44 (2), 183–211.

Velarde, R., 1905. Disponiendo se inverta la suma de 2,000 pesos en la compra de herramientas de labranza para distribuirlas entre los indígenas de la tribu Campa. *In*: C. Larrabure y Correa, ed. *Colección de leyes, decretos, resoluciones i otros documentos oficiales referentes al departamento de Loreto [1777–1908].* Vol. 5. Lima: Imp. de "La Opinión nacional,", 144–145.

Villanueva, M., 1902. *Fronteras de Loreto conferencia publica dada en la noche del 27 de diciembre de 1902.* Lima: Imprenta y Libreria de San Pedro.

Von Hassel, J., 1905. Estudio de los varaderos del Purús, Yuruá i Manu por el ingeniero Jorge M. Von Hassel. *In*: C. Larrabure y Correa, ed. *Colección de leyes, decretos, resoluciones i otros documentos oficiales referentes al departamento de Loreto [1777–1908].* Vol. 4. Lima: Imp. de "La Opinión nacional,", 209–213.

Walker, P., 2005. Political ecology: where is the ecology? *Progress in Human Geography*, 29, 73–82.

Walker, P., 2007. Political ecology: where is the politics? *Progress in Human Geography*, 31, 363–369.

Walker, R., *et al.*, 2009. Ranching and the new global range: Amazonia in the 21st century. *Geoforum*, 40, 732–745.

Walters, B. and Vayda, A., 2009. Event ecology, causal historical analysis, and human-environment research. *Annals of the Association of American Geographers*, 99 (3), 534–553.

Weiss, G., 1975. *Campa cosmology: the world of a forest tribe in South America.* New York: American Museum of Natural History.

Zimmerer, K. and Bassett, T., eds. 2003. *Political ecology: an integrative approach to geography and environment-development studies.* New York: Guilford Press.

Beyond postdevelopment: civic responses to regional integration in the Amazon

Sonja K. Pieck

Environmental Studies Program, Bates College, Lewiston, ME, USA

Postdevelopment theorists often try to read "alternatives to development" in the actions of local social movements or non-governmental organizations. More recent studies of development, however, emphasize the imbrication of state agendas with those of civil society. In the context of a major regional integration initiative in South America, this essay argues for the need to move beyond postdevelopment. Launched in 2000, IIRSA, the Initiative for the Integration of Regional Infrastructure in South America, calls for a massive expansion of the continent's transport, energy and telecommunications networks and as such has the potential to redefine both the ecological and political-economic landscapes of South America. In particular, IIRSA's Amazonian mega-projects have prompted outcries from environmentalists and indigenous peoples. In examining some of the salient methods activists are using to challenge IIRSA, this paper takes issue with postdevelopment and argues that activists are not rejecting elite projects, but instead are engaging them. Activists generate this space of engagement by invoking and problematizing the idea of active citizenship. The essay ends by exploring the implications of such a strategy, emphasizing an awareness of the entanglements between civil society and dominant institutions through a consideration of development, environment and citizenship.

Introduction

Cultural geographers have long been interested in how landscapes are imagined, by whom, and to what end (Cosgrove 1984; Rose 1993; Mitchell 1994). As indicated by the volume's overarching theme, the Amazonian landscape is a constructed and contested thing that has been shaped by the minds and hands of local and non-local groups for hundreds of years. One of the most ambitious projects to date to reshape that landscape is IIRSA, the Initiative for the Integration of Regional Infrastructure in

South America. Signed by all South American governments in 2000, IIRSA calls for a massive expansion of the continent's transport, energy and telecommunications networks through more than 500 separate infrastructure projects. By harnessing the twin powers of technology and capital, IIRSA seeks to integrate markets, improve access to resource-rich areas and create an economically productive population. Of particular interest to IIRSA's architects is the Amazon basin, and the many development projects planned here intend to transform the river, the forest, and its people into new economic opportunities. As IIRSA's neoliberal framework puts forth one set of ideas about what the Amazon means and what it should become, other visions are being offered by social justice and environmental NGOs and indigenous groups.

Through an examination of NGO responses to IIRSA's integration project, this essay argues for the need to move beyond postdevelopment. In contrast to some characterizations of postdevelopment scholars (see, for example, Sachs 1992; Rahnema and Bawtree 1997; Parajuli 1998; Leff 2005), activists are not rejecting the state or calling for the end of capitalism. Instead, they are highlighting transparency and public participation as well as adherence to the law in their effort to respond to this massive initiative. I argue that such disputes need not be oppositional in the strict sense; local groups may choose to *engage* dominant institutions by invoking and problematizing the idea of citizenship.

After situating IIRSA through a brief historical sketch of development in the Amazon, the paper discusses the limitations of postdevelopment theory. I argue that postdevelopment idealizes a complete separation of civil society from the state and that this perspective hides from our view the important debates that are crucial to so many activists in Latin America. After outlining this dilemma in more conceptual terms, this essay moves to a case study. Based on critical discourse analysis of founding IIRSA documents as well as key informant interviews with the IDB, BIC, and over a dozen of BIC's local partner NGOs active in Amazonian Ecuador and Peru in 2009 and 2010,[1] I first outline some of the salient ways in which IIRSA's creators envision regional integration, and then examine some of the methods that activists are using to challenge IIRSA, in particular their calls for active citizenship through the greater flow of information and participation in the development process. Finally, I explore the implications of activist strategies for our study of development, emphasizing an awareness of the entanglements between civil society and dominant institutions through a consideration of development, environment and citizenship.

Developing the Amazonian landscape

For hundreds of years, the Amazon has been an obstacle to, and the object of, development (Hecht and Cockburn 1990; Schmink and

Wood 1992; Cleary 2001; Little 2001; Hemming 2008; Hecht forthcoming). Amazon basin states[2] have frequently explained "underdevelopment" not only through reference to the "primitive" cultures living within their confines, but also to their expansive rainforest hinterlands (Maybury-Lewis *et al.* 2009). On the one hand, the Amazon was *terra incognita*, "unknown land" that could never be entirely understood (though efforts were taken in that regard, often by European and American naturalists). It was also a region that kept the rest of the country shackled to underdevelopment because of this (Arnold 1996; Ramos 1998; Hecht forthcoming). With time, however, the Amazon was transformed into a source of national pride and, perhaps more importantly, economic promise for many South Americans.

Military regimes and neoliberal administrations alike strongly pushed for the opening of the rainforest frontier and the development of its latent productive forces. When the continent's military regimes put forth their plans for national development, for example, in Brazil starting in the late 60s and in Ecuador in the early 70s, the Amazon played a key role. To relieve mounting tensions over land and resources in more densely settled areas, these administrations implemented broad-scale colonization programs. In a pattern that is characteristic of processes elsewhere (Wolf 1982; Blaut 1993), lands in the Amazon basin were declared "empty" despite thousands of years of indigenous habitation, and opened up to settlers who received legal title after clearing the forest and putting their new parcels to "productive" use. In other words, partially because of resource needs in other parts of the country, indigenous peoples were viewed as non-citizens, and their land uses either rendered invisible or declared worthless to the growing market economies of the region (Schmink and Wood 1984; Bunker 1985; Santos-Granero and Barclay 2000; Little 2001). In Brazil for instance, rubber barons, gold miners and ranchers all sought to profit from lands that had traditionally supported indigenous cultures. In the western headwater region of the Amazon River (in countries such as Ecuador and Venezuela), oil exploration became a driver of land-use change (Martz 1987; Coronil 1997; Gerlach 2003; Sawyer 2003).

Most recently, the Amazon basin has become the focal point of regional integration under IIRSA. IIRSA's continent-wide focus makes it the most ambitious of all development projects in the region, and its intent is to firmly connect the Amazon to the rest of South America. With the proper application of development (in this case primarily in the form of civil engineering, financed by the state and private interests), the Amazon basin can become a new opportunity for capital accumulation through export agriculture or by becoming a transit-way for goods and services from the Atlantic to the Pacific coasts. In this way, IIRSA has the potential to extend the reach of multilateral banks, country governments, and transnational capital into still marginal and resource-rich areas, literally laying the groundwork for accelerated resource extraction, export capitalism, and accumulation by dispossession (Harvey 2003) of the region's indigenous peoples. Ultimately, across Latin America today, regional integration is legitimized by a neoliberal, free-market ideology that assumes that barriers to trade stifle development (ranging from physical conditions that limit market access—like the Amazon rainforest or the Andean mountains—to tariffs or tax disincentives). Removing those barriers guarantees the free flow of goods and services that leads to economic growth and, thus, development.

The "mega-projects" that such development entails (especially highways and dams), have become targets of critique by social justice and environmental activists in the US and South America. The following section reconsiders postdevelopment theory and suggests we view activism in South America in a new light, one that permits an investigation of the complex engagements civic groups have—or seek—with dominant institutions.

Social movements and postdevelopment theory

Postdevelopment scholars identify social movements as loci for the imagination of development alternatives or even "alternatives *to* development" (Escobar 1992; see also Shiva 1989; Sachs 1992; Rahnema and Bawtree 1997; Esteva and Prakash 1998; Parajuli 1998; Leff 2005). Heavily informed by postcolonialism, poststructuralism and postmodernism, discussions of development are deeply skeptical of Enlightenment assumptions such as "progress" and "growth." The 1994 Zapatista uprising against the North American Free Trade Agreement (NAFTA), inspired scholars to celebrate the resourcefulness of local communities, stress their resistance to global capitalism, and look to (often "third world") social movements and organizations to articulate other possible worlds. For instance, in *Grassroots Postmodernism*, Esteva and Prakash consider the "peoples beyond modernity" and highlight the ways in which their wisdom offers an alternative to capitalism's "Goliath" (1998, pp. 4–6). Anthropologist Arturo Escobar has perhaps the most clearly articulated argument in this respect (1992, 1995, 2001). For him, "[s]ocial movements constitute an analytical and political terrain in which the weakening of development and the displacement of certain categories of modernity (for example, progress and the economy) can be defined and explored" (1992, p. 28). In contrast to the decontextualizing forces of capitalism (Harvey 1996; Lefevbre 1991), place-based (but not place-bound) social movements thus become the site of new imaginations for human-environment relations based on principles of justice, autonomy and care.

Postdevelopment literature has contributed to a re-appreciation of marginalized communities, has attempted to extend the critique of development, and has raised important questions about its driving assumptions and ultimate desirability. Yet postdevelopment has attracted criticism from various corners for "populist myopia" (Watts 2003, p. 28), and for creating all too neat binaries between the West and non-West (Corbridge 1998). Civil society does not stand in opposition to development; it is, in fact, often entangled with it. Instead of postdevelopment's much lauded "resistance" (which is read as a clear rejection of development and the search for separate alternatives), social movements or NGO activism may engage with dominant groups. For example, in her study of Afro-Colombian movements, Kiran Asher (2009, p. 25) argues that:

> these communities, and the social movements that emerge from them, are not simply manifestations of radical, non-Western culture, as post-developmentalist discourses of difference imply. Rather, interconnections between structural forces and divergent discourses (including those of nation-state formation, or, more recently, globalization and environment-alism) are important elements in understanding how local identities and interests are shaped and articulated.

In other words, social movements and development processes are some-times mutually constitutive rather than necessarily oppositional. As social movements engage with powerful institutions, we cannot always simply read this as a rejection of elite power; we need to re-introduce the idea of entanglement with dominant institutions such as the state or multilateral development banks (Pieck and Moog 2009). As Asher observes: "[i]n demanding autonomy *from* the state, the PCN [Proceso de Comunidades Negras, the Afro-Colombian Movement] recognizes the state's authority and in a curious way legitimizes it by engaging its instruments (laws, policies) and institutions" (2009, p. 128, original emphasis; see also Li 2007). And in his work on slum dwellers' construction of urban peripheries in Brazil, Holston sees "insurgent citizenship." He explains this analytical choice:

> I do not study these developments, as some theorists and activists do by separating civil society and state. Nor do I view the mobilization of social movements as the resistance of the former and their demobilization as cooptation by the latter. I avoid such dichotomies by focusing on citizenship as a relation of state and society, and I study its processes to reveal the entanglements of the two that motivate social movements to emerge and subside. I examine these processes as they appear in the practices of citizens. (Holston 2008, p. 9, 13)

While postdevelopment theorists have written eloquently about the environmental and human costs of development (Shiva 1989; Leff 2005), and the "alternatives *to* development" offered by local communities or local NGOs, the perspectives opened up by more recent scholarship can deepen our understanding of the far-ranging debates around development. If activists are actually *engaging* elite projects rather than rejecting them, then struggles around development projects can speak to issues beyond ecological and social sustainability. While a conflict is on the surface about the protection of environments and lifeways, the dispute may also raise fundamental questions about the relationship between communities, the state and capital, and the meaning of citizenship.

History and purpose of IIRSA

IIRSA entered into force after a meeting of all South American heads of state in August 2000 in Brasília, Brazil. The meeting sought to "advance the modernization of regional infrastructure through the adoption of specific actions to promote its integration and socio-economic development" (IIRSA 2004, p. 3). A "Plan of Action" was drawn up during a December 2000 meeting of transportation ministers and has served as a point of reference for the development of projects. Repeated meetings have occurred once or twice a year since IIRSA's launch.

On paper, IIRSA's key rationale has been explained as follows:

> In the current context of globalization, the main challenge for the first decade of the new millennium is to secure a higher rate of sustained growth. This should spring from productive processes based on technology and knowledge, and ever less from dependence on the exploitation of natural resources. Only growth based on real increases in productivity and competitiveness can make the region internationally relevant, while simultaneously creating the conditions for a pattern of sustainable development that is stable, efficient and equitable. (IIRSA Technical Coordinating Committee 2000, p. 1)

Some authors also point to larger economic trends that help us understand the reasoning behind IIRSA, emphasizing the growing significance of Asian markets—especially China—and their role as trading partners (Mesquita Moreira 2006; VanDijck 2006; Shadlen 2008). IIRSA documents promise that the initiative will integrate South America, establish the infrastructure necessary to transport goods and facilitate communication between markets, and thus unleash the economic potential of the continent's resources (the region boasts ample supplies of some of the world's foremost factors of production: oil, minerals, genes/biodiversity, and water). In my conversation with IDB staff, however, additional rationales emerged. First, IIRSA exists to economically connect countries previously torn apart by conflict. Second, IIRSA seeks to create these connections through the extraction of raw materials and their export to key markets abroad, thus entrenching a primary commodity economy.

Third, IIRSA also serves the purpose of developing previously peripheral, non-market areas:

> [T]he idea is to have more integrated countries in South America, to have more connection. You need infrastructure, especially [between] Brazil [and] the other countries. Imagine, the history of these countries is a history of disintegration; it's not a history of integration.
>
> And there is another thing that I think is a factor of change. It's to integrate the 'hinterland' of South America. Because you know, South America on the Brazilian side is well-developed on the coast, and on the other side, our neighbors in Peru, Ecuador on the Pacific [are also well developed]. IIRSA is a way to help the hinterland to start to be developed. (IDB staff member, interview with author, 16 April 2010)

The integration of the "hinterland" into the continent's economy thus connects closely with earlier projects to link resource-rich areas to the capital cities. Undergirding ideas like this is the assumption that without that connection, those areas continue to be unproductive backwaters that are home to equally unproductive people—wasted economic potential. Staff members at the IDB confirmed that the strongest advocate for IIRSA, and its founding visionary, was the Cardoso government of Brazil. As the continent's economic powerhouse, Brazil presented and has advanced IIRSA in an attempt to secure its claim on resources and connection to markets (IDB staff member, interview with author, 16 April 2010; Camacho and Molina 2005).

Implementing IIRSA

Under IIRSA, South America has been partitioned along ten "axes of integration and development," also referred to as "hubs" in English (IIRSA Technical Coordinating Committee 2000; IDB 2004; IIRSA 2004). These are the: (1) Andean hub; (2) Guyana shield hub; (3) Amazon hub; (4) Peru-Brazil-Bolivia hub; (5) Central interoceanic hub; (6) Capricorn hub; (7) Mercosur-Chile hub; (8) Southern hub; (9) Paraguay-Paraná waterway hub; and the (10) Southern Andean hub (see Figure 1).

These lines of demarcation are based on strategic factors, such as key waterways, or other shared geographical characteristics. Overall, IIRSA's ten hubs encompass close to 50 project groups that include 519 individual projects (McElhinny and Velásquez-Donaldson 2009), approximately 40% of which are under construction, 10% have been already completed, and another 20% are currently in the planning stages (Laats 2009). This brings the total estimated investment so far to $55.6 billion (McElhinny and Velásquez-Donaldson 2009).

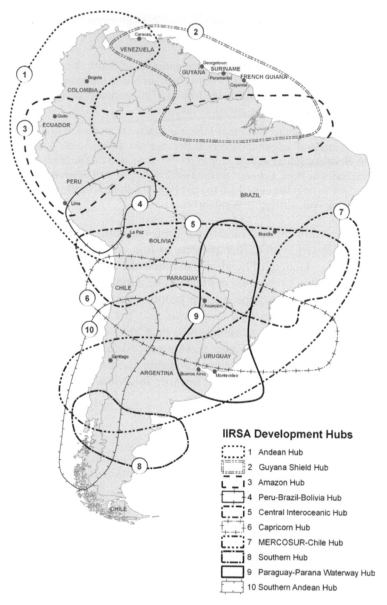

Figure 1. IIRSA's ten Integration and Development Hubs. (Adapted from: IIRSA 2004, p. 17.) Source: Used with permission from Inter-American Development Bank.

To get a sense of the complexity and ambitiousness of IIRSA, the Amazon hub alone covers four different countries (Ecuador, Colombia, Brazil and Peru), stretches across 4.5 million square kilometers that are home to 52 million people, and consists of seven project clusters. One such cluster alone consists of nine projects, each of which could be a hydroelectric dam or a section of road (Marcondes-Rodriguez 2008, p. 17). The financing bodies for IIRSA are the Inter-American Development Bank (IDB), the Andean Development Corporation (CAF), and the Fund for the Development of the Río Plata Basin (FONPLATA).[3]

In sum, implementation of IIRSA will require vast expenditures of funds from multilateral, corporate and governmental sources, draw on the technological capacities of private contractors and state agencies, and the changes it unleashes will reverberate throughout the continent.

IIRSA's ecological and social impacts in the Amazon

IIRSA's development projects, particularly the roads, pipelines, and dams, will have major ecological and social impacts if completed. According to a recent study conducted by Conservation International, IIRSA's projects, most crucially its expansion of roads, will further propel the agricultural frontier into forested areas (Killeen 2007; Laurance *et al.* 2001; Nepstad *et al.* 2006). Of primary concern in Peru, for instance, is the Interoceanic Highway that runs from Rio Branco in Brazil to Puerto Maldonado in southern Peru. From there the road cuts through the Andes to Cusco and Arequipa and then to various ports on the Pacific coast. The road has already increased deforestation and has led to an expansion of illegal mining activity in the area (Dourojeanni *et al.* 2010). Other concerns center on the effects of increased in-migration and colonization by landless settlers (Hecht and Cockburn 1990; Schmink and Wood 1992; Fearnside and de Alencastro Graça 2006; Mendoza *et al.* 2007; Killeen 2007; ProNaturaleza, TNC, and WWF staff members, interview with author, 9 and 17 June 2010).

Increased infrastructure development in the Amazon lowlands will also make oil and gas reserves more accessible through pipeline construction. As experience with the Brazil-Bolivia, Camisea (Peru) and OCP (Ecuador) pipelines has shown, these infrastructure projects have "spurred further exploration and production, because once the transport problem is resolved, additional upstream investments are necessary to fill the pipelines" (Killeen 2007, p. 30). Industrial accidents, oil spills and inadequate treatment of wastes have contributed to environmental deterioration and public health catastrophes in places experiencing increased development such as northeastern Ecuador (Kimerling 1990). Much of the wealth generated by oil companies does not directly benefit the population most affected by this industry. As a result, hydrocarbon production has emerged as one of the most contested issues in South America. As Killeen notes, opposition to multinational oil companies has fueled social unrest and even "has contributed to the success of recent political candidates in Ecuador and Bolivia" (2007, p. 31). Finally, IIRSA foresees the construction of numerous hydroelectric dams along the

Amazon River and its tributaries. Some of these, such as the vast Rio Madeira (Brazil) or Iñambari (Peru) dam complexes, are either planned, already financed, or under construction and are causing concern on the part of environmentalists, and social justice and indigenous rights activists (Killeen 2007, pp. 37–38). As is already known from similar projects in Brazil and beyond, hydroelectric dams disturb normal river flow, create ecological changes upstream (where large areas are flooded) and down-stream (where areas may desiccate), can lead to the extinction of terrestrial and aquatic species, and jeopardize the livelihoods of the region's population, including indigenous communities (Fearnside 2006; International Rivers 2008).

Years after IIRSA's launch, a loose network of civil society organizations began forming to contest and redefine a number of IIRSA's key propositions. In the process, these groups are articulating new discourses around human-environment relations that reorient nature and citizenship away from IIRSA's economic logic.

Civil society's response to IIRSA in the Amazon

As this volume emphasizes, landscapes are products of physical and social processes. In the context of IIRSA, the Amazon basin has become yet again a site for the contestation of multiple visions. IIRSA's many mega-projects, and especially those in sensitive ecological areas such as the Amazon basin, have prompted outcries from civil society groups such as environmentalists, social justice activists and indigenous peoples. By charging that environmental and social impact assessments are unsatisfactory, that the implementation process is nontransparent and oppressive, and that alternative territorial imaginations and lifeways are being silenced, environmental and indigenous rights groups seek to alter the terms on which IIRSA extends and legitimizes its control over Amazonian lifeworlds.

In the following sections, I describe several ways in which civil society actors attempt to disrupt IIRSA's logic. What becomes clear, however, is that this is not a case of outright rejection of development or the search for "alternatives *to* development." Activist networks are not "displacing categories of modernity" like "progress" and "the economy" (Escobar 1992, p. 28). Rather, in their engagement with IIRSA and its advocates, activists are spending their energies trying to make room for public participation in modernity.

Creating multiscalar networks

One of the major platforms against IIRSA has formed among northern NGOs, spearheaded by the Bank Information Center (BIC). With funding from the Moore Foundation,[4] BIC has developed a major civil society initiative called BICECA (Building Informed Civic Engagement for

Conservation in the Andes-Amazon). The goals of BICECA are to help "civil society organizations to analyze and influence economic integration projects and policies in the Andes-Amazon in order to help protect the biological and cultural diversity of the region" and to "promote informed engagement and effective conservation advocacy through linking civil society initiatives in the local, regional, and international arenas" (BIC 2009).

The primary outgrowth of this initiative is the generation of a transnational activist network. BIC hosted a meeting of northern and southern civil society organizations in Lima, Peru in July 2005, which culminated in the *Articulación Frente a IIRSA* (Platform against IIRSA). At this meeting, the *Articulación* drafted its founding document, the Lima Declaration, in which it defines its goals as:

> unmasking and halting the IIRSA initiative in the way it is currently being implemented, thus contributing to the political and social construction of a critical consciousness concerning IIRSA, and building society's capacity for intervention, and to generate alternative sustainable processes in order to achieve another possible form of integration. (Lima Declaration 2005)

The Lima Declaration was signed by sixteen NGOs, including BIC.[5] Since 2005, over forty groups have joined the *Articulación*. There are three clusters of organizations that now comprise this network: (1) US-based internationally active environmental and social justice NGOs (e.g., the Center for International Environmental Law and BIC, both in Washington, D.C.); (2) South American environmental and social justice NGOs located in the capital cities (e.g., Asociación Civil Labor in Lima, Peru or Ecolex in Quito, Ecuador); and (3) indigenous federations in the Amazonian lowlands.

Reframing sustainability

In the IIRSA context, while documents outwardly state a vague commitment to "sustainable territorial development" (IIRSA Technical Coordinating Committee 2000; IDB 2004; IIRSA 2004), the principal thrust of the initiative is geared towards resource extraction, the integration of markets (both regionally and internationally) and increased productivity and competitiveness through infrastructure projects and institutional changes. The network of NGOs and grassroots groups, however, is engaging in a redefinition of sustainability that puts environmental health and social justice front and center. In the Lima Declaration, the civil society organizations declare that:

> We support South American integration within a concept that respects our peoples, their cultures, and the environment, and that at the same time promotes locally supported development. We reiterate that government

actions should support truly sustainable development and should seek to improve the living conditions of the populations living in those ecosystems, which for centuries have conserved and managed their natural resources in a sustainable manner. Our territories, with their rivers and forests, are a vital part of the complexity of life on our planet, which is being profoundly threatened by the large dams, industrial waterways, and inter-oceanic corridors that are part of IIRSA. (Lima Declaration 2005)

According to the *Articulación*, "truly sustainable development" includes a respect for local people's livelihoods, the myriad social and environmental relations that constitute places, and an awareness of how global processes (like regional integration and through it greater insertion into a global economy) can produce negative environmental and social change (Lima Declaration 2005). Such sustainable development, however, must come about through "government actions," not the lack thereof.

Such statements are critiques of the "field of intervention"—in other words, the framing of the problem and its implied solution—that states and development banks have articulated. As Tania Li (2007, p. 18) argues, "governmental interventions can never achieve all they seek. An important reason promised improvements are not delivered is that the diagnosis is incomplete.... It cannot be complete if key political-economic processes are excluded from the bounded, knowable, technical domain." Social movements and NGOs have recognized the problematic assumptions that legitimize IIRSA's form, and have begun identifying the failures and gaps in development that are caused by the neglect of local voices. Social movements and NGOs thus politicize place and social relations, demanding that IIRSA, and the governments that drive it, incorporate and thus make visible that which they have so far obscured. This may mean that "government actions" that "improve the living conditions of the populations living in those ecosystems" are warranted, as stated in the Lima Declaration passage quoted above. In other words, as the *Articulación* begins to reframe sustainability, we simultaneously find, contrary to postdevelopment, an appeal to the state.

Questioning IIRSA's governance structures

One of the key demands emanating from social movements and NGOs is for the IIRSA planning and implementation processes to include the full and active participation of civil society. The Lima Declaration states that "a different form of integration is possible, but only through a broad, informed debate, with the full participation of all of our peoples" (Lima Declaration 2005). Civic groups consistently point to the failure of IIRSA governance structures to be transparent and fair. According to a document released by BIC, IIRSA lacks a "formal mechanism to promote civil society participation in project identification, evaluation, and execution" (BIC 2004). In other words, groups are demanding an institutionalized space for the grassroots in the governance structures of IIRSA. Only the inclusion of a "broad range of actors" can provide the "crucial input" (BIC 2004) that will help prevent ecological and social damage. "Participation" in this sense should occur at all stages of the planning process, including the decisions over what the regional integration priorities should be in the first place. Put differently, participation connotes a search for power within, rather than a wresting of power from, dominant institutions.

Another point of contention is that IIRSA lacks transparency. NGOs seek "timely access to unambiguous information" that allows for "meaningful participation from civil society" (BIC 2004, p. 5). This would include information on how decisions are made, how integration priorities are set, how anchor projects are identified, and what forms private sector involvement takes in each project. Information, however, has not been forthcoming. In November 2008, twenty civil society organizations from eight South American countries requested a meeting with IIRSA directors to discuss how to increase civil society participation in the IIRSA planning process. The request was rejected (BIC 2008a). And IIRSA's website, the primary method for official dissemination of information about the initiative, has failed to provide information to civic actors about how decisions are made, how IIRSA projects are planned and implemented, and how exactly they are funded (BIC staff person, interview with author, 8 July 2009). And of course, not all grassroots actors have access to computers or the internet in the first place (BIC 2004). IIRSA may thereby offer transparency without granting it to certain clearly defined subsections, for example, those without access to computers or the internet, notably those living in rural rainforest communities.

These demands are significant because they, also, challenge IIRSA's forms of rule and control. One of the ways power is secured and maintained by institutions is to hide its workings from the view of (potentially antagonistic) others (Weber 1978). In this way, power can be "spatialized" and comes to be seen as dominant and acting "from above" and outside the control of local actors (Scott 1998; Ferguson and Gupta 2002). By calling for an opening up of the governance spaces that have remained obscure, this activist network is also insisting on shining a light on the techniques of regulation and rule in hopes of making them visible and hence amenable to grassroots influence. Civil society groups insist on having increased power in the decision-making process and this network, in fact, challenges the naturalized dominance of IIRSA's strategies and its assumed spatial verticality.

One of the successes of this network is the fact that the Corporación Andina de Fomento (CAF), one of IIRSA's financing bodies and, according to BIC, one of the least transparent organizations of IIRSA, recently agreed to release documents regarding funding flows to various

IIRSA projects (BIC 2007) so that civil society groups could see the private donors and understand some of the decision-making behind the scenes. Another accomplishment of the *Articulación* was a 2008 meeting hosted by IIRSA's technical coordinating committee about its Amazon projects. The meeting not only brought together IIRSA's experts, but also assembled various corporate actors such as Odebrecht (engineering and petrochemicals), Andrade Gutierrez (construction) and Petrobras (energy), as well as a number of civil society organizations like the World Wildlife Fund-Brazil and BIC "to debate the compatibility of large, high risk infrastructure projects in the Amazon with sustainable, equitable development outcomes" (BIC 2008b). While BIC considered the meeting modest in relation to the magnitude of issues requiring discussion, the action nonetheless speaks to an opening of IIRSA's governance and administrative structures.

Beyond postdevelopment: articulations of active citizenship

In Latin America, the notion of citizenship is an especially contested one. It has functioned for a number of social movements and grassroots actors as a vehicle for forging common ground and for demanding and expanding social and political rights during Latin American's transition from authoritarian to democratic rule in the 1980s (Eckstein 2001; Yashar 2005; Dagnino 2006). Significantly, the struggle for citizenship in the region entailed not only the redefinition of existing rights but the creation of new ones (including the right of autonomy over one's own body and the right to a safe environment) (Dagnino 2006, p. 25). For indigenous and black movements at this time, the rights to land and defense of territory—as linchpins of self-determination and cultural survival— emerged as key ingredients to citizenship (Ramos 1998; Brysk 2000; Yashar 2005).

Scholars note, however, that the meanings of citizenship are subtly changing under neoliberalism. For Dagnino (2006), the collective meaning attributed to citizenship by social movements is being replaced by an individualistic, atomized understanding. Neoliberalism also establishes "an alluring connection between citizenship and the market. To be a citizen is the individual integration into the market, as a consumer and as a producer" (2006, p. 40). Along these lines, Yashar (2005) discusses the paradox of neoliberal citizenship: the expansion of certain (individual) political rights concurrent to the denial of (often collective) rights to subsistence resources. This is particularly troubling in that "the recognition of rights seen in the recent past as an indicator of modernity is becoming a symbol of backwardness"—the market thus gets re-envisioned as a "surrogate site for citizenship" (Dagnino 2006, p. 41; Gledhill 2005; Valdivia 2005; Taylor 2009).

As a result, the notion of political "participation," a hard-fought concession during the 1980s and 1990s—has been resignified. Across Latin America, various forms of neoliberalism have led to a closing of the more politicized spaces that were won with great effort by social movements during the 1980s and 1990s. Among these are the elimination of social and labor rights in Brazil (Dagnino 2006, p. 41), the violent repression of resistance to extractive industry in the Amazon basin (as happened in Peru in 2009) or, in Guatemala, the emergence of neoliberal multiculturalism. This has opened some space for cultural politics, but enshrines particular cultural rights only insofar as those rights do not fundamentally challenge state and capital (Hale 2002). Not surprisingly, IIRSA's sense of citizenship aligns with these ideas.

This "shrinking of citizenship," from the more politicized spaces of the 1980s to a narrower, "sterilized" version that equates participation with individual integration into the market economy, is evident in the IIRSA discourse. Under IIRSA, the population is seen as a latent source of productivity. For instance, one IDB presentation described South America's "young, entrepreneurial population" and its "dynamic, creative cultures" as "an opportunity" (IDB 2004). In other places, the population is described as not having access to markets, thus implying that its productive potential (much like the region's natural resources) has yet to be tapped (IDB 2006). Under IIRSA, "empowerment . . . is the harnessing of the poor to help remove barriers to their participation in market relations" (Taylor 2009, p. 35). In addition, while IIRSA documents seem to agree in principle on the value of civil society participation, actual commitments (e.g., in the form of meetings both at the IDB and at the country level, or by providing access to financial documents) have not been realized.

The struggle around IIRSA is a struggle around the meanings and the practice of citizenship in South America. Mechanisms for civil society participation in IIRSA, as discussed above, are limited and perfunctory at best. In some ways, citizenship then emerges as a rather passive act, a matter of becoming a "productive" individual integrated into the market economy and, through self-betterment, contributing to the good of the nation. For the NGO and grassroots members of the *Articulación*, however, citizenship is active. The work of BIC and its local partners centers on two elements: information dissemination and capacity-building. In the context of IIRSA, being a citizen means in part contributing to the definition of development, redefining it, recrafting development priorities and having full involvement in the identification of projects and decision-making capacity in both planning and implementation stages. Crucially, this includes access to information and the ability to process that information and use it to one's benefit. This is the impulse behind the *Articulación's* online "knowledge bank" (BIC 2009). The information on the BIC website originates from diverse sources, both from the BIC staff, but importantly also from BIC's many regional partner organizations in Amazon basin countries (BIC staff member, interview with author, 8 July 2009). In BIC's view, the lack of knowledge about IIRSA leaves people powerless to respond and unable to hold IIRSA's institutions accountable for negative social and ecological impacts that are already occurring in the wake of various high-impact projects. BIC has recently expanded tactics, from information politics at the transnational level to advocacy through educating and mobilizing people on the ground and working more closely with local partners to do so (BIC staff member, interview with author, 8 July 2009).

In contrast to postdevelopment perspectives, BIC's local partners do not want to do away with either the development project or the state that drives it. In my conversations with NGO staff, one of the points of agreement was to avoid a rejection of development and instead make sure that development proceeded in the "right" way:

> Our organization doesn't have a position either in favor or against the [Inter-Oceanic Highway in southern Peru]. Infrastructure projects can be done, but with better mitigation standards. So the road is not intrinsically "bad," but it exists in a structural context of poverty. And that creates long-term, negative environmental impacts [through in-migration and settlement]. And that needs to be recognized. (ProNaturaleza, Peru, interview with author, 15 June 2010)

> We don't want to reject infrastructure development wholesale, we just want to make sure that it goes well. (Forum Solidaridad, Peru, interview with author, 16 June 2010)

> I mean, if you ask me, I think integration is a fabulous idea. It just needs to happen with the appropriate standards. (Sociedad Peruana de Derecho y Ambiente, interview with author, 17 June 2010)

Another principal point of agreement for those NGOs whose work is directly political (as opposed to conservation organizations whose work is not) was the need for the state to provide the "right" kind of development and the lack of information that does not permit the average citizen to make precisely that demand. As one staff member at Peru's Instituto del Bien Común stated with evident frustration: "How can you insist on your rights when you don't have any information [about the development project]?" (interview with author, 14 June 2010).

Knowledge (of IIRSA) is power (the capacity to voice concerns, to demand change, or to protest). For the NGOs, while it constitutes a strategy to avoid a public outcry or further strain an already tense relationship with the citizenry, the fact that governments continue to withhold information from the public is a failure of governance. A staff member at the Asociación Civil Labor office in Peru explained the demand for information this way: "Because of IIRSA, Peru is in debt to the Inter-American Development Bank. That's public debt (deuda pública). That makes it *our* issue" (interview with author, 15 June 2010). Citizenry and government are bound together by the law and constitution—a social contract. When governments remain quiet about massive development projects like the Interoceanic Highway, or fail to conduct legally required environmental impact assessments, they put themselves above that contract and the accountability that should come with it. In response to my question about what they considered to be currently the greatest challenge for them regarding IIRSA, staff members at the Ecolex office in Ecuador exclaimed: "We have to somehow sensitize our government to its own norms. The government *itself* must follow the law" (interview with author, 9 June 2010).

Rather than seeing the regulatory apparatus of the state as a threat, it becomes the fallback, the guarantor of something better. There is still faith that the letter of the law, of the constitution even, can be a source of redress, a way to tame the state's development excesses and ensure "truly sustainable development" in line with strict environmental and social standards. It is up to civil society (in this case, NGOs) to monitor this, to get the word out to the public, and to develop their capacity to insist on their rights. In the words of a staff member at Forum Solidaridad in Peru, the group attempts to disseminate information about IIRSA projects so that local communities can "advocate for themselves" (interview with author, 16 June 2010).

Both the BIC strategies and the kinds of responses I received from NGO activists are not indications of an emerging postdevelopment agenda. We are not witnessing "the politics of 'No'"(Esteva and Prakash 1998, p. 32). Instead, what we find is an engagement *with* the state, a desire for the state to be more participatory, more transparent. Throughout all my conversations, the development project itself (and underlying notions of progress) was not questioned. What did incur frustration, however, was states' heavy, exclusive focus on "desarrollo de asfalto" ("asphalt development," staff member, ProNaturaleza, interview with author, 15 June 2010) without appropriate safeguards, without adherence to the governments' own stated commitments, and without a more holistic notion of progress.

As postcolonial states become "governance states" (Harrison 2004)—states whose governance structures are being complemented by a host of other international institutions—the locus of accountability risks being shifted farther away from local communities and towards international institutions like the IDB or less transparent funding bodies like CAF or FONPLATA. Ultimately, the reaction to IIRSA echoes the anxieties of other movements across the hemisphere, like the Zapatista uprising against free trade, the Bolivian water and gas wars, or the landless movements in Brazil. Here as well, activists and community members demand greater accountability of their governments at a time when neoliberal economic policies threaten to further displace already tenuous links between states and societies.

Yet we should also remain aware that reactions differ across space, conflict and groups. While "another world is possible" for the Zapatistas in Mexico, activists in other places put forth other ideas. In the case of IIRSA and the *Articulación* the response revolves around enabling the practice of democratic citizenship in the face of expanding capitalism. The key word here is "attempting"—we should not fall into the romanticization of postdevelopmentalism. For instance, NGOs in countries like Peru (where one of IIRSA's most controversial projects, the Interoceanic Highway, is nearing completion) face an uphill battle. Relationships to the central government are tense, in part due to ex-president Fujimori's authoritarian style, and current president García's violent repression of protest.[6] In contrast to the deeply entrenched oil and mining industry, environmentalism has not long been institutionalized in the country (staff member, Instituto del Bien Común, interview with author, 14 June 2010). Coalition-work like the kind described above has produced results (like the increased flow of information to the IIRSA website, a meeting in Brazil on Amazon projects or the release of funding information on the part of CAF), but overall, Peruvian activism has not yet led to a shift in decision-making processes.

Conclusion

As this volume makes clear, Amazonia is both a material and imaginary space. For hundreds of years, this region has been linked to the global market. Over the past decades especially, the Amazon basin has seen rapid changes related to development. For Amazon basin states, the forest has filled ambiguous roles, at times a wasteland and at others a treasure trove of unrealized opportunity. Under IIRSA, the Amazon is both: with the help of infrastructure technology, geographic barriers like the forest or the river can either be eliminated or used to economic advantage, and thus the region's promise will finally be fulfilled. Yet as the other authors in this collection of papers also argue, the ever-changing Amazon once again gives birth to new physical, social and metaphorical landscapes, and new identities (see also Cosgrove 1984; Mitchell 1994; Slater 2002). In this case, IIRSA's plans have prompted diverse populations to articulate different understandings of this space Importantly, social justice and environmental activists do not reject development, as suggested by postdevelopment scholars. Instead, local and transnational groups attempt to re-orient development towards particular civic needs and thus call on governments and the Inter-American Development Bank (and the capitalist interests they often represent) to permit the participation of those most affected by IIRSA.

Landscapes are sites of cultural and political contestation and in the process of this struggle new imaginings of place are generated and circulated. Some of them might be hybrids of grassroots and elite worlds. Our studies of development in the Amazon as both idea and practice, must consider the possibility that activists do not, in fact, reject development. Instead, struggles around development that occur at the crucible of state, capital and civil society often raise thorny issues about the nature of the relationship between these actors. Through the idea of citizenship, the tensions between dominant institutions and subaltern communities emerge as the contested terrain, where new relationships between states, societies and their environments are being forged in South America.

Acknowledgements

The author would like to thank the editors, the article reviewers, and the individuals who kindly agreed to be interviewed for this study. Additional thanks to the Bates College Imaging Center for its assistance developing the IIRSA map.

Notes

1. Indigenous groups sometimes have a different platform than their NGO allies, and this is to some extent the case with IIRSA. Similarly, the struggle around regional integration involves many additional stakeholders such as workers, traditional peoples, or rubber tappers. This essay focuses on environmental and social justice NGOs only. Understanding the positions of other stakeholders on IIRSA is part of necessary future research.
2. The Amazon basin stretches across nine different South American countries: Bolivia, Brazil, Colombia, Ecuador, French Guyana, Guyana, Peru, Suriname and Venezuela.
3. IIRSA was signed in 2000, that is, before a number of leftist-populist administrations came to power in South America (e.g., Lula in Brazil, Correa in Ecuador, and Morales in Bolivia). When I spoke with officials in the Inter-American Development Bank, I was told that despite the ideological shift, all governments continue to be actively engaged in pushing IIRSA forward, while attempting to negotiate the most favorable parameters for themselves. One of the reasons for this is that IIRSA is focused exclusively on the technical aspect of infrastructure, rather than political institution-building or legal changes.
4. The three-year grant was approved in March 2005. The Moore Foundation's description of the purpose of this grant states: "This grant supports the development of an information infrastructure and the exchange of information about the Bank Information Center's economic integration initiative in South America.... Outcomes include design and implementation of a user friendly technology platform for information storage and dissemination, five to seven case studies detailing the impact of international financial institutions in the Andes-Amazon region, creation of an information network to communicate about IIRSA, alignment of civil society actors, creation of a collective vision for

conservation development, and proposed mitigations to policymakers" (Moore Foundation 2009).

5. A complete list of signatory organizations is available here: http://www.bicusa.org/en/Region.KeyIssues.100.aspx [accessed 17 February 2009].

6. In June 2009, indigenous people in Peru's Bagua province took to the streets to protest a new land reform law favoring foreign investors. The Peruvian government responded by sending troops to Bagua and shooting into the crowd from helicopters. The clash killed 11 police officers and a still unknown number of indigenous civilians.

References

Arnold, D., 1996. *The problem of nature: environment, culture and European expansion.* Oxford: Blackwell.

Asher, K., 2009. *Black and green: Afro-Colombians, development, and nature in the Pacific lowlands.* Durham, NC: Duke University Press.

Bank Information Center (BIC), 2004. *A preliminary critique of IIRSA* [online]. Available from: http://www.bicusa.org/bicusa/issues/IIRSA%20piece.pdf [Accessed 1 March 2009].

Bank Information Center (BIC), 2007. *La CAF responde la solicitud de información de parte del BIC* [online]. Available from: http://www.bicusa.org/es/Article.3373.aspx [Accessed 1 March 2009].

Bank Information Center (BIC), 2008a. *Directores de IIRSA cierran espacio para participación de la sociedad civil* [online]. Available from: http://www.bicusa.org/es/Article.10936.aspx [Accessed 29 May 2009].

Bank Information Center (BIC), 2008b. *IIRSA CCT meeting on infrastructure and the Amazon: IIRSA's future in doubt* [online]. Available from: http://www.bicusa.org/en/Article.3907.aspx [Accessed 1 March 2009].

Bank Information Center (BIC), 2009. *The BICECA project: a response by civil society to regional integration* [online]. Available from: http://www.bicusa.org/en/Region.100.aspx [Accessed 22 July 2009].

Blaut, J.M., 1993. *The colonizer's model of the world: geographical diffusionism and Eurocentric history.* New York: Guilford Press.

Brysk, A., 2000. *From tribal village to global village: Indian rights and international relations in Latin America.* Stanford University Press.

Bunker, S.G., 1985. *Underdeveloping the Amazon: extraction, unequal exchange, and the failure of the modern state.* The University of Chicago Press.

Camacho, G.H. and Molina, S., 2005. IIRSA y la integración regional. *Observatorio social de América Latina,* 17, 307–316.

Cleary, D., 2001. Towards an environmental history of the Amazon: from prehistory to the nineteenth century. *Latin American Research Review,* 36 (2), 65–96.

Corbridge, S., 1998. 'Beneath the pavement only soil': the poverty of post-development. *Journal of Development Studies,* 34 (6), 138–148.

Coronil, F., 1997. *The magical state: nature, money and modernity in Venezuela.* University of Chicago Press.

Cosgrove, D., 1984. *Social formation and symbolic landscape.* Madison, WI: University of Wisconsin Press.

Dagnino, E., 2006. Meanings of citizenship in Latin America. *Canadian Journal of Latin American and Caribbean Studies*, 31 (62), 15–52.

Dourojeanni, M., Barandiarán, A., and Dourojeanni, D., 2010. *Amazonía peruana en 2021*. 2nd ed. Lima, Peru: Sociedad Peruana de Derecho y Ambiente.

Eckstein, S., ed. 2001. *Power and popular protest: Latin American social movements*. Berkeley, CA: University of California Press.

Escobar, A., 1992. Imagining a post-development era? Critical thought, development and social movements. *Social Text*, 31–32, 20–56.

Escobar, A., 1995. *Encountering development: the making and unmaking of the third world*. Princeton University Press.

Escobar, A., 2001. Culture sits in places: reflections on globalism and subaltern strategies of localization. *Political Geography*, 20 (2), 139–174.

Esteva, G. and Prakash, M.S., 1998. *Grassroots postmodernism: remaking the soil of cultures*. London: Zed Books.

Fearnside, P.M., 2006. Dams in the Amazon: Belo Monte and Brazil's hydroelectric development of the Xingu river basin. *Environmental Management*, 38 (1), 16–27.

Fearnside, P.M. and de Alencastro Graça, P.M.L., 2006. BR–319: Brazil's Manaus-Porto Velho highway and the potential impact of linking the arc of deforestation to central Amazonia. *Environmental Management*, 38 (5), 705–716.

Ferguson, J. and Gupta, A., 2002. Spatializing states: toward an ethnography of neoliberal governmentality. *American Ethnologist*, 29 (4), 981–1002.

Gerlach, A., 2003. *Indians, oil, and politics: a recent history of Ecuador*. Wilmington, DE: Scholarly Resources.

Gledhill, J., 2005. Citizenship and the social geography of deep neo-liberalization. *Anthropologica*, 47 (1), 81–100.

Hale, C., 2002. Does multiculturalism menace? Governance, cultural rights and the politics of identity in Guatemala. *Journal of Latin American Studies*, 34, 485–524.

Harrison, G., 2004. *The World Bank and Africa: the construction of governance states*. London: Routledge.

Harvey, D., 1996. *Justice, nature and the geography of difference*. Malden, MA: Wiley-Blackwell.

Harvey, D., 2003. *The new imperialism*. Oxford University Press.

Hecht, S. and Cockburn, A., 1990. *The fate of the forest: developers, destroyers and defenders of the Amazon*. New York: Harper Collins.

Hecht, S.B., forthcoming. *The scramble for the Amazon: imperial contests and the tropical odyssey of Euclides da Cunha*. Chicago: University of Chicago Press.

Hemming, J., 2008. *Tree of rivers: the story of the Amazon*. London: Thames & Hudson.

Holston, J., 2008. *Insurgent citizenship: disjunctions of democracy and modernity in Brazil*. Princeton University Press.

IIRSA, 2004. *IIRSA project portfolio* [online]. Available from: http://www.iirsa.org/BancoConocimiento/L/lb04_cartera_de_proyectos_iirsa_2004/lb04_cartera_de_proyectos_iirsa_2004.asp?CodIdioma=ENG [Accessed 20 February 2009].

IIRSA Technical Coordinating Committee, 2000. Action plan for regional infrastructure integration in South America. *Meeting of the Ministers of*

Transport, Telecommunications and Energy in South America, Montevideo, Uruguay, 4–5 December 2000 [online]. Available from: http://www.iirsa.org [Accessed 19 February 2009].

Inter-American Development Bank (IDB), 2004. *The role of the Inter-American Development Bank and civil society in the integration process* [online]. Available from: http://www.iadb.org/sds/doc/sgc-iirsaandcivil_society.pdf [Accessed 15 January 2010].

Inter-American Development Bank (IDB), 2006. *Building a new continent: a regional approach to strengthening South American infrastructure* [online]. Available from: http://idbdocs.iadb.org/wsdocs/getdocument.aspx?docnum =834660 [Accessed 27 December 2008].

International Rivers, 2008. *Águas turvas: alertas sobre as conseqüências de barrar o maior afluente do Amazonas* [online]. Available from: http:// internationalrivers.org/en/am%C3%A9rica-latina/os-rios-da-amaz%C3%B4nia/ rio-madeira/%C3%A1guas-turvas-alertas-sobre-conseq%C3%BC%C3% AAncias-de-barrar-o- [Accessed 18 July 2009].

Killeen, T., 2007. A perfect storm in the Amazon wilderness: development and conservation in the context of the Initiative for the Integration of the Regional Infrastructure of South America (IIRSA). *Advances in Applied Biodiversity Science* 7 [online]. Available from: http://library.conservation.org/portal/ server.pt/gateway/PTARGS_0_122814_129379_0_0_18/AABS.7_Perfect_Storm _English.low.res.pdf [Accessed 30 May, 2009].

Kimerling, J., 1990. *Amazon crude.* New York: Natural Resources Defense Council.

Laats, H., 2009. IIRSA y UNASUR. *World Social Forum*, Belém, Brazil, 27 January–1 February 2009 [online]. Available from: http://www.bicusa.org/ admin/Document.100727.aspx [Accessed 1 March 2009].

Laurance, W.F., *et al.*, 2001. The future of the Brazilian Amazon. *Science*, 291 (5503), 438–439.

Lefebvre, H., 1991. *The production of space.* Malden, MA: Blackwell.

Leff, E., 2005. La geopolítica de la biodiversidad y el desarrollo sustentable. *Observatorio Social de América Latina*, 4 (17), 262–273.

Li, T.M., 2007. *The will to improve: governmentality, development, and the practice of politics.* Durham, NC: Duke University Press.

Lima Declaration, 2005. [online] Available from http://www.bicusa.org/Legacy/ FinalversionDeclaration%20of%20Lima%20English.pdf [Accessed 26 February 2009].

Little, P.E., 2001. *Amazonia: territorial struggles on perennial frontiers.* Baltimore, MD: Johns Hopkins University Press.

Marcondes-Rodriguez, M., 2008. *Myths and reality about IIRSA.* Presentation of the Inter-American Development Bank at the Wilson Center, 16 January 2008 [online]. Available from: http://www.wilsoncenter.org/events/docs/mauro. marcondes.iirsa.pdf [Accessed 2 December 2008].

Martz, J., 1987. *Politics and petroleum in Ecuador.* New Brunswick: Transaction Books.

Maybury-Lewis, D., MacDonald, T., and Maybury-Lewis, B., 2009. *Manifest destinies and indigenous peoples.* New York: David Rockefeller Center for Latin American Studies.

McElhinny, V. and Velásquez-Donaldson, C., 2009. Visiones de integración regional: lecciones aprendidas. *World Social Forum*, Belém, Brazil, 27 January–1 February 2009 [online]. Available from: http://www.bicusa.org/admin/Document.100725.aspx [Accessed 1 March 2009].

Mendoza, E., *et al.*, 2007. Participatory stakeholder workshops to mitigate impacts of road paving in the southwestern Amazon. *Conservation and Society*, 5 (3), 382–407.

Mesquita Moreira, M., 2006. *IIRSA economic fundamentals* [online]. Inter-American Development Bank. Available from: http://idbdocs.iadb.org/wsdocs/getdocument.aspx?docnum=800737 [Accessed 6 December 2008].

Mitchell, W.J.T., 1994. *Landscape and power*. University of Chicago Press.

Moore Foundation. 2009. Grants awarded: Bank Information Center (BIC). Available from: http://www.moore.org/grant.aspx?id=670, accessed 23 June 2009 [Accessed 23 June 2009].

Nepstad, D., Stickler, C.M., and Almeida, O.T., 2006. Globalization of the Amazon beef and soy industries: opportunities for conservation. *Conservation Biology*, 20 (6), 1595–1603.

Parajuli., P., 1998. Beyond capitalized nature: ecological ethnicity as an arena of conflict in the regime of globalization. *Cultural Geographies*, 5, 186–217.

Pieck, S.K. and Moog, S.A., 2009. Competing entanglements in the struggle to save the Amazon: the shifting terrain of transnational civil society. *Political Geography*, 28 (7), 416–425.

Rahnema, M. and Bawtree, V., eds. 1997. *The post-development reader*. London: Zed Books.

Ramos, A.R., 1998. *Indigenism: ethnic politics in Brazil*. Madison, WI: University of Wisconsin Press.

Rose, G., 1993. *Feminism and geography: the limits of geographical knowledge*. Minneapolis, MN: University of Minnesota Press.

Sachs, W., 1992. *The development dictionary: a guide to development as power*. London: Zed Books.

Santos-Granero, F. and Barklay, F., 2000. *Tamed frontiers: economy, society, and Civil rights in upper Amazonia*. Boulder, CO: Westview Press.

Sawyer, S., 2003. *Crude chronicles: indigenous politics, multinational oil, and neoliberalism in Ecuador*. Durham, NC: Duke University Press.

Schmink, M. and Wood, C.H., 1992. *Contested frontiers in Amazonia*. New York: Columbia University Press.

Schmink, M. and Wood C., eds. 1984. *Frontier expansion in Amazonia*. Gainesville, FL: University of Florida Press.

Scott, J., 1998. *Seeing like a state: how certain schemes to improve the human condition have failed*. New Haven, CT: Yale University Press.

Shadlen, K., 2008. Globalisation, power and integration: the political economy of regional and bilateral trade agreements in the Americas. *Journal of Development Studies*, 44 (1), 1–20.

Shiva, V., 1989. *Staying alive: women, ecology, and development*. London: Zed Books.

Slater, C., 2002. *Entangled Edens: visions of the Amazon*. Berkeley, CA: University of California Press.

Taylor, M., 2009. The contradictions and transformations of neoliberalism in Latin America. *In*: L. Macdonald and A. Ruckert, eds. *Post-neoliberalism in the Americas*. London: Palgrave, 21–36.

VanDijck, P., 2006. Infrastructure for deeper integration in South America: IIRSA. *Meeting of the Latin American Studies Association*, 15–18 March 2006 San Juan, Puerto Rico.

Valdivia, G., 2005. On indigeneity, change, and representation in the northeastern Ecuadorian Amazon. *Environment and Planning A*, 37 (2), 285–303.

Watts, M., 2003. Development and governmentality. *Singapore Journal of Tropical Geography*, 24 (1), 6–34.

Weber, M., 1978. *Economy and society*. Berkeley, CA: University of California Press.

Wolf, E., 1982. *Europe and the people without history*. Berkeley, CA: University of California Press.

Yashar, D., 2005. *Contesting citizenship in Latin America: the rise of indigenous movements and the postliberal challenge*. Cambridge: Cambridge University Press.

The new Amazon geographies: insurgent citizenship, "Amazon Nation" and the politics of environmentalisms

Susanna B. Hecht

School of Public Affairs, Institute of the Environment, Department of Geography, UCLA, Los Angeles, CA, USA

This article reviews the main themes—we are here; this is who we are; insurgent citizenship; Amazon Nation—elaborated in this collection of papers on new Amazonian geographies, and extends their implications to ideas about governmentality and regional identity. The article contextualizes the papers in this issue through understanding Amazonia's role in the structuring of the contemporary Brazilian state through resistance to conventional modernist authoritarian development planning, and the creation of current places and politics through the assertion of new forms of citizenship, identity, governance and the rise of socio-environmentalisms as part of a new "statecraft" from below. Modern Amazonia has reasserted itself by developing a set of alternative epistemes and practices which can be seen, in their language and ideologies, to invoke the idea of the "Amazon nation." This article emphasizes the cultural underpinnings of these processes in a contested Amazon that is now a major supplier of global agricultural commodities in its deforested landscapes, and pivotal for local livelihoods and planetary environmental services in its forested ones.

Introduction

Amazonia, largely viewed as a torpid backwater at mid-20th century and a place whose prominence seemed well in the past, emerged startlingly into the Brazilian forefront when, in 1964, the military seized power. As a central piece of authoritarian nation-building, the territorial integration of Amazonia was to become a key policy of the new regime. As the generals gazed north and prepared for Amazonian occupation, they viewed the region as largely empty, a *tabula rasa*, on which to inscribe their ambitions (Becker 1982; Mattos 1980; Silva 2003). Its river capitals with their crumbling opera houses and decaying palaces were pathetic reminders of an earlier glory, and embarrassing proofs of the meager developmentalist capacities of Amazonian inhabitants.

Like much of Latin America in the post-World War II period, it was thought that development interventions would whisk these countries from colonial backwardness through the stages of economic growth into a sleek Euro-American modernism (Rostow 1971; Holston 1989; Schelling 2001; Scott 1998). According to the new military masters of Brazil, Amazonia, with its shabby river towns and vast forests, needed a modernist, technical and technocratic vision, one that would transform the inchoate emptiness with its smattering of pointless Indians, and its revanchist extractive economy, into a theater of modernist statecraft. (Silva 1967; Martins and Zirker 2000; Birkner 2002). If Brasília was the urban manifestation of modernist development, Amazonia would be its rural exemplar. By applying the organizing frameworks of natural resource assessment, the ideas of French regional planner François Perroux's version of growth poles and development corridors, and the finest of Brazil's modern entrepreneurs, the wilderness would shift from a "land without history" into the dizzying landscapes of "Manifest Destiny" or as the Brazilians called it, the "March to the West" (See Garfield 2001; Becker 1982). This type of integration had actually been a central idea in Brazil's territorial imaginary since the 18th century, and a dream of the Brazilian Republic since its inception in 1889 (Hecht 2011b). Later it was articulated by President Getulio Vargas in the 1930s as a policy of nationalist development when he invoked the "vast and fertile valleys of the west" that would construct the "instruments of our defense and industrial progress"(Vargas 1938, p. 125). The idea was expanded during the authoritarian regimes begun in 1964, and inspired the geopolitical doctrines of military theorist Golbery de Couto e Silva, who viewed national integration, and especially that of Amazonia, as a central national security issue both in its development and military dimensions (Silva 1967; Mattos 1980; Becker 1982; Birkner 2002). The 1964 military government would transform this "vast unwanted present"—the demographic void of the Amazon—by means of a radically different future scripted out in its master National Development Plans.[1] Lands were gridded out, straight roads were punched through the jungle, development growth pole maps were drawn, and fiscal incentives proffered within the framework of growth pole planning. Meanwhile, colonization programs, whose purposes were to reduce pressure for agrarian reform elsewhere, were put into place. Unsurprisingly, the region experienced an accelerated dynamic of deforestation associated with roads, land claiming, speculation, ranches and farms, as the new forms of state legibility—derived from radar and satellite imagery, to the angular lines of rectangular pastures or "fishbone" patterns of colonization at the household plot level—filled remote sensing images of Amazonia. As in earlier territorial "integrationist" moments of Amazonia's histories, from the first European contact, through the progressive Pombaline era,[2] to the rubber boom (Amazonia's modern integration into the global systems), processes were unleashed that involved massive population dislocations, new administrative systems, new tenurial regimes, restructuring of the economy, integration into new markets, and significant environmental changes (Nugent 1993; Cleary 2001; Raffles and WinklerPrins 2003; Harris 2010).

As the roads unfurled, there was an unsettling discovery: the idea of empty Amazonia was a myth, and Amazonians actually had histories and historic claims over their landscapes. The region was soon embroiled in what looked like a land war, but was also a more profound "clash of cultures," an unanticipated assertion of alternative livelihoods, economic models and futures. What was opening up as the trees fell were the spaces of what Holston and Appadurai call "insurgent citizenship" where "invisible" cultures and populations began to affirm their claims as *citizens* (Holston and Appadurai 1996).

This collection of papers on emerging Amazonian geographies takes this as its point of departure. The ambitious landscapes of modernist planning and technocratic development—the interoceanic highways, Initiative for Integration of Regional Infrastructure in South America (IIRSA), the planned construction of more than 149 dams, the neoliberal frontiers of soy rotations stretching from Maranhão to Santa Cruz, Bolivia, that use deforested lands as a substrate for introduced, often exotic production systems[3]—remain in contest with the worlds of socio-environmentalisms (*socio-ambientalismo*). These socio-environmentalisms viewed nature and regional societies as co-produced over time, through daily practices, politics and through differing epistemes of nature from the modernist land as substrate point of view. These are not conceptualized in an environmentally deterministic way, but are highly dynamic. Their "construction" is better understood as framed by political ecology rather than North American environmentalism, which historically has concentrated on conservation in "wild" landscapes.[4] These "insurgent" Amazonian societies and their socio-natures articulate a different version of the future, and they must adapt to and reshape the institutions and practices of conventional infrastructure and regional planning, even as the latter must adapt to regional politics. What lies at the nexus between these increasingly globalized modernities is the environmental question, where socio-environmentalisms currently hold the trump cards because of the pivotal role tropical forests play in biodiversity and climate change.

Context: from the margins of history to its center

The globalization of environmental concerns as a popular cause rather than the grumblings of a few ecologists, began in Amazonia with the extension of the military's authoritarian development model in the 1970s, and the subsequent regional and international uproar over deforestation.

Simple conservation had never been part of the Brazilian political landscape, where environmental complaints tended toward urban pollution and urban amenities rather than the global scope that was rapidly defining Amazonian development politics (Foresta 1991). Even more disconcerting to the military regime at the height of the dictatorship were the alliances of Brazilian counterparts with international groups engaged in human rights, environmental justice and critiques of ecological destruction (Hecht and Cockburn 1989; Anderson *et al.* 1991; Schmink and Wood 1992; Wood and Schmink 1993; Hochstetler and Keck 2007). There were many examples of such alliances, including those focused on controversies over indigenous policies as well as traditional peoples, like rubber tappers and other forest extractivists. These Amazonian battles marked off an entirely new form of globalized politics that extended from humble Amazon backwaters to the most august chambers of the World Bank, all condemning the authoritarian modernist enterprise in Amazonia as ecologically catastrophic, socially unjust, and economically corrupt.

On the ground, forest populations regularly faced off against developers and an indifferent state. Such battles were rife in the Amazon in the 1970s and 80s, and were generally resolved through the standard violence and impunity. In the forests of the Western Amazon one victim, however, was not shunted into the unmarked grave and oblivion usually reserved for those who ran up against rancher elites. Chico Mendes' martyrdom thrust Acre into the glare of global scrutiny and public outrage. Mendes, who had been a rural labor organizer and Workers' Party activist, was more engaged in the concerns of social justice than environmentalism, but at this time, in this part of the Amazon, these efforts merged (Hecht and Cockburn 1989; Mendes *et al.* 1992; Rodrigues and Rabben 2007). Overnight, Mendes became an Amazonian eco-icon, a symbol of populist nature preservation pitted against elite developer greed. In the wake of Mendes' assassination, the Brazilian government and the murderers themselves were stunned by the international outcry over what they regarded as the very ordinary termination of an obscure labor leader. What they were encountering was the first globalized tropical conservation movement allied to powerful national associations ranging from the Forest People's Alliance, the Workers' Party and Brazilian human rights organizations, to international environmental organizations like the Environmental Defense Fund, and Amnesty International, all clamoring for a different tropical development politics (Hecht and Cockburn 1989; Schmink and Wood 1992; Nugent 1993; Harris and Nugent 2004).

Within this history, the Mendes' legacy is complicated and interesting, and speaks to the ways that his dreams of social justice, forest life, woodland economies and political regimes were recast through the way the larger world came to perceive him, and how the politics of tropical development were changed through his martyrdom. The extractive reserve, an inhabited forest, called into question both the imperious Generals' "modernist development" approach that depended on deforestation, and the model of empty parks promoted by international conservation organizations. This "all or nothing" view of Amazonian possibilities was previously taken as the only imaginable set of choices—choices that left little room in either case for local Amazonian inhabitants. As Brazilian politics opened, the idea of an inhabited forest as a viable element of regional development, with new tenurial structures and different natural resources institutions with a broader mandate than simple forestry, materialized in the form of a vast network of extractive reserves and extensive indigenous territories. As well as being part of a local livelihood strategy, these reserves became part of a defensive regional strategy used by communities to block forest clearing, as deforestation frontiers moved further into the region. Research increasingly showed that inhabited forests were far more effective than parks for curbing clearing (Pedlowski et al. 2005; Merry et al. 2006; Nepstad et al. 2006a).

With the rewriting of the Brazilian Constitution in 1988, and the ratification of Article 68, which affirmed the land rights of traditional peoples, a powerful legal framework was put into place for the development of a range of institutions that could support the populations whose claims to Amazon landscapes had largely been denied. The historic exclusion of most Brazilians from property ownership, especially in Amazonia, had made "illegality" the norm, and this illegality denied most poorer Amazonians political rights. People were thus vulnerable to wholesale land usurpations of areas they had occupied for generations, as the "misrule" of law obfuscated, deflected, and intimidated the poor, while those with power traditionally ignored or manipulated the legal formalities. (Davis 1977; Kotscho 1981; Holston 1991; Nugent 1993; Alston et al. 1999; Hemming 2003; Harris and Nugent 2004; Rival 2004). What Article 68 did was transform people once classified as "squatters" into "citizens" if they could articulate a politics of history of place more powerful than the strategies of simple fraudulent land grabs that had become so characteristic of Amazonian occupation (Branford and Glock 1985; Hecht and Cockburn 1989; Schmink and Wood 1992; Alston et al. 1999; Fearnside 2007; Simmons et al. 2007; Merry et al. 2008; Hecht 2010). Thus, "territorialization" in the Amazon land debates often required insurgent citizenship, resurgent identities (as in the case of CITA—Conselho Indigena do Tapajós-Arapiuns, see Bolaños 2011, and also Porro et al. 2011), and arguments about forest stewardship (see Minzenberg and Wallace 2011) and over what constitutes a forest (see Vadjunec et al. 2011, and Mentore 2011)—in order to counter competing land claims. The new possibilities of citizenship through the struggles over landscapes and territory almost by definition required the assertion of forest cultures and identities, and moved inhabitants of Brazil's most peripheral universes into legitimated political contests with national and international powers.

Four themes inspiring speculations about new Amazonian geographies

The papers in this collection suggest four central themes in the current literature that integrate Amazonian people, environment and politics. They do so in ways very different from the simple microeconomic commodity calculus that is marshaled to mobilize funding for the new mega-projects that are progressively moving off the drawing boards into the landscape. This section outlines these themes and processes that now shape the new Amazonian geographies.

"We are here ..."

The first main theme is that the Amazon is not empty, and it hasn't been empty for at least 10,000 years. This of course seems like a given, yet the idea of the "demographic void" still persists and the invisibility of traditional populations of all kinds throughout Amazonia underpins so much of its history of violent conflict. The premises and practices of the authoritarian planners, developers and classic preservationists were often unraveled by previously unseen populations, who, as the political context changed, changed the rules of the game by asserting historical claims and refusing to leave. Traditional people increasingly shape regional politics and the political ecologies of development. If there were a single factor that one could identify that most shattered preconceived ideas of Amazonia, it would be the idea of human presence. Invoking historical documents, historical geographies and local knowledge systems, land and territorial claims are made on the basis of human rights, historical ecologies inscribed in landscapes, traditional tenurial regimes, and continuous occupation. The fact of still "being there" after centuries of attempts to enslave and control these populations speaks to Amazonians' formidable powers of resistance (Hemming 1987; Harris 2010).

"This is who we are ..."

Who exactly these people were and are suggests the next general theme: identity and territory. This gives rise to questions of "authenticity"—are people what they say they are?—and a significant politics of reinscription of identity, a dynamic seen especially in the case of the Council of the Lower Tapajós-Arapiuns, and the Babassu extractors as the papers by Bolaños (2011) and Porro *et al.*, (2011) show so well. This discovery and reclassification of populations is a major feature of Brazilian land politics everywhere, as emergent ethnicities and previous *quilombolas* (members of former slave communities) asserted their claims over historically, ecologically and discursively constructed landscapes (Restall 2005; French 2006; Treccani 2007; Perz *et al.*, 2008; Hecht 2010). The complexity of ethnicity as a historical and contemporary outcome of politics is also revealed in the study of the Asháninka presented in the paper by Salisbury *et al.* This unshackling of the complex ethnogenic history of Amazonia away from the simple rubric of "peasants" and "detribalized" Indians has resulted in the discovery of much more complex social histories, different policies and practices (See Bolaños 2011; Porro *et al.* 2011; Salisbury *et al.* 2011; Vadjunec *et al.*, 2011), a position at odds with both the conventional and Marxist development literatures that viewed rural peoples as more or less undifferentiated classes to be dealt with within a simplified modernizing framework, rather than disparate, contentious cultural entities operating in particular ecologies (Nygren 2000; Brass 2002, 2005; Martins 2002; Bicker *et al.*, 2003; Otero and Jugenitz 2003; Reed 2003; Rival 2003; Harris and Nugent 2004; Hecht 2010).

The combination of "presence" and the assertion of identities and rights have produced a new kind of Amazonian citizenship associated with the defense of territories and lifeways. A range of institutional arrangements now build on existing practices as legitimate development alternatives to formal development projects and agro-industrial development with their long history of environmental destruction. Thus, as Vadjunec *et al.*, (2011) show, the idea of a "forest government" using a forest-based development model clearly expresses a concrete development alternative. Laura Mentore's (2011) discussion of the Waiwai within British Guyana's Low Carbon development model, under the hoped-for flows of REDD (Reduced Emissions from Deforestation and Degradation), points out the competing ideas of nature that will have to be negotiated—including the meaning of forests and who has sovereignty over them—in the upcoming land politics of carbon trade.

Insurgent citizens

The third theme is that of citizenship. The role of rural socio-environmental movements in the struggles for democratic opening, their links to national labor movements, and casting environmental concerns as a defense of livelihood, has given Amazonians an unusual symbolic role and significance as a practical model in Brazilian land politics. More notably social movements have helped cast the ideas of environmental justice well beyond the frameworks of exposure to pollution, or differential access to environmental amenities, or expressions of race-based privilege, to assert that struggles over natural resources are simultaneously struggles for social justice and nature. (For another context see Perrault 2008.) What is clear is that Amazonian development politics are now largely inconceivable without addressing socio-environmental movements and the competing debates over land use and resources. The effective struggles over land and the evolution of new institutions produced a politics of alternative development approaches, procedures, and cultural survival.

What is unfolding in Amazonia is a new kind of region-making, based on new kinds of social morphologies, informed by different world views that are derived from life and practices. Most recently these have been structured by a range of interacting resources, national regional and international markets and evolving practices in politics, institutions and production activities (Hall 2000; Harris and Nugent 2004; Sears *et al.,* 2007; Brondizio 2008; Padoch *et al.*, 2008). This idea of "becoming" part of broader, distinctive regional polity, through which one engages one's rights through politics and social engagement is a feature of a great deal of historical ethnography, and though little studied in Amazonia, it remains a fertile research field as the papers in this collection surely show.

This citizenship in the "Amazon nation" unfolds in a national context but is negotiated through the maze of Amazonian institutions and ideologies, rather different from those of the nation at large, and plays out at local levels through the architectures of local power. In this light, the actual workings of citizenship in Acre through *Florestanía* (Vadjunec *et al.,* 2011), through negotiations and exchanges over hunting in sustainable development reserves (Minzenberg and Wallace 2011), the *articulación* of social movements and affected populations of IIRSA projects (Pieck 2011), the politics of the Babassu women (Porro *et al.*, 2011), and the resurgent identity politics of CITA (Bolaños 2011), all reveal in different ways how different places have shaped the meaning of belonging and self-understanding in complex, distinctly Amazonian ways. They form part, in other words, of an imagined community, to use Anderson's term (Anderson 1991). Florencia Mallon's study of peasantries in the making of nations points out how peasant communities mobilized by external threats, and embracing a liberal discourse about equality and citizenship, engaged in political processes and cultures (and guerilla war) to produce what Mallon calls "alternative nationalisms" (Mallon 1995). For the insurgents, this idea of imagined community, being a nation, grew in relation to its opposite: being a colony (Mallon 1995), and was a way of constructing a new type of more inclusive nationalism and autonomy. While clearly the forest states of Acre, Amazonas and Amapá are Brazilian, they manage to advance a quite different version of regional identity and development compared to those in southern Brazil.

Amazon Nation

While there is national content in the framing of citizenship as the authors of these papers describe it, what seems at least as important is regional identity, the "Amazon Nation" which has distinctive histories and trajectories compared with more general emblems of Brazilian-ness, Peruvian-ness, and Guyanan coastal culture, although clearly the issue is most developed for Brazil. Since the First Republic, Brazil has used the Europeanized southern version of itself as its emblematic projection on the international stage, with its nature and natives seen as folkloric "add-ons." The Amazon and its peoples represented a nostalgic and exotic past now transcended in modern Brazil. Indeed, the bustling megalopolis of São Paulo and the industrial south reveal how much Brazil seems to have transformed itself. But the questions about the long-term resilience of this model are now part of a broader global questioning. Amazonian approaches to citizenship reject a homogenizing frame of national identity—the usual feature of nationalism—in favor of difference, new social morphologies, and regional ideas in development politics (Holston 2008). The *Florestania* of the forest government of Acre invokes the classic ideological elements of nationalism: a shared history, struggles, suffering and triumph, a spatial configuration of lifeways and "our habits and our ways." The forest states of Acre, Amazonas, and Amapá all engage a language of historical difference, popular citizenship, and distinctive socio-and ecological landscapes to counter the conventional development discourses and practices that are largely predicated on forest removal. All have responded with concrete proposals ranging from forest-based commodities to environmental services, rooted in community management in interaction with the local states. By doing so they hybridized global and local ideologies of forest-based sustainability, but within Amazonian worldviews and identity. This thinking in a "regional way" highlights grounded political strategies that are quite different from the earlier practices of resistance: *Nuestra Bien Viver*, *Florestania*, and ethnopolitical action by CITA and the Waiwai in Guyana take on a politics of articulation, using this term in both its meanings of expression and connection. The "territorialization" of identity movements in Amazonia has scaled up into a regional one, as they more or less democratized Amazonian space and aspects of its planning.[5] Thus for Amazonians, their livelihoods and histories are not seen as some atavistic residue of an unfortunate history, but the social, intellectual and ecological foundations of development alternatives (Hall 1997; Perreault 2003; Gomes 2005; Campos and Nepstad 2006; Lima *et al*, 2006; Rodrigues and Rabben 2007; Brondizio 2008; Segebart 2008; Viana 2009; Hecht 2011a). If the modernist project and economic integration of Amazonia represented a new iteration in nation making from southern Brazil, Amazonian "alternative nationalisms," it could be argued, constitutes another.

The idea of "Amazon Nation" is helpful for overcoming the intellectual conundrum that Amazonia is simultaneously "Big Nature" and extremely urban. National identities are not limited to particular economic livelihoods, so whether urban Acrianos are currently tappers is not relevant in the current sense of Amazonian cultural belonging, any more than being a Jeffersonian yeoman farmer is a prerequisite for today's American identity. Amazonian urban dwellers have an extremely complex economic relation with forests, as urban analysts are discovering, revealing webs of connections as anastomose as the region's landscapes. The widespread existence of multi-sited households, and the complex bricolage between urban and rural social networks and economies, has a venerable history in Amazonia (see, for example, Nugent 1993; Raffles 2002; Harris 2010). This is just as true today. Such a complex network of activities, economic strategies, social relations, and markets is found from one end of the Basin to the other, and in its complex mobilities, the categories of urban and rural are much more permeable than the temperate zone ideas of the urban, and elsewhere in Brazil. Global and local markets, remittances, pensions, schooling, and household subsidies of many kinds flow in both rural to urban and urban to rural directions, and embrace fluidity, perhaps the best over-arching metaphor for Amazonia (Padoch 1999; Padoch *et al.*, 2008; Brondizio *et al.*, 2009; Brondizio *et al.*, 2011; Pinedo-Vasquez and Padoch 2009; Parry *et al.*, 2010; Perz *et al.*, 2010; Padoch *et al.*, 2011).

The project on New Social Cartographies of the Amazon, coordinated by anthropologist Alfredo Wagner and Rosa Azevedo Marin, executed with Ford Foundation funding in association with many grassroots political organizations, shows the complexity of "hidden" livelihoods and spatial organization, making possible a review of social, economic and ecological situations through which livelihoods and natures) are produced.[6] The emphasis on bringing to light previously invisible systems helps make their resource and economic claims visible to the state, and these systems also become inputs in alternative economic proposals. This brings us to another realm where Amazonian cultural and political ecologies are having an impact of contemporary social theory. This is in the sphere of governmentality.

Governmentality, environmentality and the statecraft of the small

This brings me to the final question, which pertains to the idea of governmentality or because it addresses the environment, what Arun Agrawal (2005) has called "environmentality." This Foucauldian term is used to discuss how governments manage the "conduct of conduct" and it refers less to the "hard" controls of the Panopticon of "discipline and punish" (or more popularly "command and control") than to the soft controls of the integration of regulatory thoughts and norms into the very integument of daily institutions, practices and subjectivities (Agrawal 2005). It refers to how one experiences and views the world, why one "cares about" things and then aligns one's behavior. This is the action of "intimate" government. The discussions of governmentality in general, and environmentality in particular, focus on decentralization policy and the practices of community in governance of natural resources—the technologies of the state through which environmental rules are now formally enacted—and how they literally become "embodied" in the subjectivities and values of the local population, and then constructed through the practices and local political ecologies to produce forests. This is a "co-construction," or co-evolution of place with people. Agrawal's dazzling discussion of how a population in Kumaon India, came to care about forests through "intimate government" is a landmark ethnography about the internalization of environmental concerns with the expansion of conservationist ideologies in state practices.

This strikes me as an inappropriate model for Amazonia. As Mentore (2011) shows so clearly, as do other authors in this collection and a raft of other Amazonian ethnographic studies (see Hill 1988; Moran 1993; Balée 1994; Block 1994; Descola 1994, 1996; Porro 1996; Wright 1998; Fausto 2001; Albert and Ramos 2002; Rival 2002; Posey and Plenderleith 2004; Fausto and Heckenberger 2007), there is clearly a different set of epistemes—world views—of nature that underpin Amazonian "subjectivities." In fact, at the end of the day it has been environmental pressure from Amazon communities—allied to more formal national, global and regional environmentalisms—that has shaped Amazonian statecraft and ultimately Brazilian environmental politics and institutions. A little more than a decade ago the governors of Amazonas, Mato Grosso, and Pará preened in their resistance to international outcry over forest clearing, and were largely indifferent to the concerns and formal critiques about the development model ripping apart their landscapes. Today state governors ranging from soy magnate Blairo Maggi to green governor Eduardo Braga, (and indeed the President of Guyana) view the natures of Amazonia as part of their global competitive advantage. They assert that the vocation for Amazonia is woodlands, especially in light of climate change and the possibility of Reduced Emissions from Deforestation and Degradation (REDD) funding as part of the global politics and economies of climate change, as developed countries invest in Amazonian forests as carbon sinks.[7]

Ever since the death of Chico Mendes and the rise of the Forest People's Alliance, Amazonian "environmentality" has expressed the "taming power of the small" (to quote the *I Ching*) as the Amazonian experience began to fundamentally shape the politics of Brazil, "greening" its national discourses about development and changing the relevant institutions, at least in so far as they apply in Amazonia (Brown and Rosendo 2000; Anderson and Clay 2002; Kainer *et al.*, 2003; Goeschl and Igliori 2004; Dandy 2005; Ruiz-Perez *et al.*, 2005; Hochstetler and Keck 2007). The "environmentality" in Amazonian politics is not so much a

213

function of state decentralization and urgings that were transposed into "intimate government" of subjects, but rather the inverse. The epistemes, ideologies and practices of stewardship of many Amazonian cultures—their own internal "intimate governments" and subjectivities, their identities, communities and ecologies—have transformed the institutions and practices of the local and national state, opening up the way for alternatives of many kinds, even though the region still remains highly contested.

Nation building "from below" has been a vibrant topic of Latin American historians interested in the practices and politics of early nation construction elsewhere in Latin America (Joseph and Nugent 1994; Mallon 1995; Chasteen and Castro-Klarén 2003; Chernilo 2007). They are at pains to point out that the processes of political construction and nation building are continuous political tasks. What have largely gone unnoticed are the profoundly new forms of "Amazon Nation" that now unfold in post-authoritarian Amazonia. One might argue that Amazonian environmentality (and its kindred institutions and financiers) has been so successful that it has now profoundly transformed regional modernist political ecologies: since 2004, the Amazonian rate of deforestation has declined by 70% (INPE and Espacias 2010). As Pieck (2011) points out, constant vigilance and politics underpin such gains, and politics is by definition processual and there is much to be gained and lost in Amazonia.

Final thoughts

The papers in this volume show how fruitful Amazonia remains for complex interdisciplinary studies, and how relevant much of the discussion of cultural histories and political ecologies can be for analyzing planning and development politics, and for imagining sustainable and resilient trajectories. In this concluding essay I have tried to integrate the implications of these case studies into a larger regional framework examining how these "add up" and contribute to larger debates about regions, nations and governance. Amazonia remains exceedingly dynamic, and although there are superb case studies of natives and traditional peoples, what the region rather lacks now are studies of how the "taming powers of the Great"—that is how large powerful groups often associated with the energy and mining economies—are unfolding. There have been a few recent studies on soy (Jepson et al., 2005; Hecht 2005; Jepson 2006; Nepstad et al., 2006b; Hecht and Mann 2008; Brannstrom 2009), and the impacts of the livestock sector are again under review (Toni 2007; Michalski et al., 2008; Walker et al., 2009), but the central dynamics of these systems are still somewhat open to question. Mining has all but disappeared from the literature, and the discussions of forestry remain largely focused on the ecology of this sector's impacts (Stoian 2005; Sears *et al.*, 2007; Bowman *et al.*, 2008; Amacher *et al.*, 2009; Fearnside 2008; Medina, *et al.*, 2009a, b; Hoch *et al.*, 2009). The larger questions of the how the interactions of the energy sector—including biofuels, fossil fuels, and hydropower—will shape the region's future remain interrogations. The great question for Amazonia at this time is the effect of REDD in the region, and the dynamics of its political economies, especially as this new form of economic input faces off with that of massive infrastructural development.

The great Brazilian writer and Amazon explorer, Euclides da Cunha felt that Amazonians, his "bronzed titans," held the key to success in the region, and counterposed their land husbandry to those aspirants who had had "apprenticeships in plunder" (Hecht 2011b). The papers in this volume show why that might be so, as well as the complexity and incomplete natures of these alternative proposals. The gains made by defenders of the region are important, because, as da Cunha also noted, Amazonia was (and still is) the "last unwritten page of Genesis."[8] Amazonia however, is full of Black Swans, to use Taleb's phrase for unpredictable events that have very large impacts (Taleb 2009). Most of the transformative events there were largely unpredictable for the powers of the time: the destruction of natives by European disease, the Cabanagem revolt of the 19th century, the rise and fall of the rubber economy, and the emergence of socio-environmentalisms, all transformed the region in profound and unexpected ways. The future is likely to be no different. In this once again, Da Cunha must have the last word: "Such is the river, such is its history: always turbulent, always insurgent, always incomplete."[9]

Notes

1. See Holston (1989, 1999) for discussions about Brazilian modernist planning.
2. This refers to the period when the Sebastião José de Carvalho e Melo, the Marquis of Pombal, radically restructured Portuguese national and colonial policy. He implemented a state charter company (Compania de Grão Para e Maranhão) to organize the trade, subsidized the importation of slaves with which he hoped to overcome the labor scarcity, and exiled religious orders whom, as a sharp anti-cleric, he viewed as monopolizing indigenous labor. Thus began the Directorate which replaced religious orders with secular directors who had power over native labor, beginning another period of major indigenous dislocation He invested in major military forts and infrastructure, and sought to "modernize" Amazonian economies, within the frameworks of the time, through expanding plantation agriculture and colonization (Hecht 2011b; see also Maxwell 1995, 2001).
3. "Exotic" is used here in the sense of non-native or introduced species. Thus pastures depend on cattle introduced either from Asia (Zebu), Africa (Creole) or Mediterranean breeds, on grasses largely initially imported from Africa.

Soy is an Asian import; sunflower is part of the soy rotation and is from North America.
4. The papers in this volume point to these differences. Other recent commentators include: Turner and Robbins 2008, Peres and Zimmerman 2001, and Zimmerer 2000, 2006.
5. In this regard, see also Escobar (2008).
6. Details on this project can be obtained from Ford Foundation Brazil: www.fordfoundation.org/regions/brazil. See also: www.novacartografiassocial.com/archovos/pnca.
7. The discussions of REDD are too extensive to address here, but see Borner and Wunder (2008), Borner and Wunder (2010), Chhatre and Agrawal (2009), Grainger et al., (2009) Sandbrook et al., (2010), and Sikor et al., (2010).
8. See Hecht (2011b). Translation from the Portuguese "terra sem historia: observações gerais." (Tocantins 1966, p. 32)
9. See Hecht 2011b.

References

Agrawal, A., 2005. Environmentality-community, intimate government, and the making of environmental subjects in Kumaon, India. *Current Anthropology*, 46, 161–190.

Albert, B. and Ramos, A.R., 2002. *Pacificando o branco: cosmologias do contato no Norte-Amazônico*. São Paulo: Editora UNESP.

Alston, L.J., Libecap, G.D., and Mueller, B., 1999. *Titles, conflict, and land use: the development of property rights and land reform on the Brazilian Amazon frontier*. Ann Arbor: University of Michigan Press.

Amacher, G.S., Merry, F.D., and Bowman, M.S., 2009. Smallholder timber sale decisions on the Amazon frontier. *Ecological Economics*, 68, 1787–1796.

Anderson, A.B. and Clay, J.W., 2002. *Esverdeando a Amazônia: comunidades e empresas em busca de práticas para negócios sustentáveis. Brasília. DF* São Paulo: IIEB Fundação Peirópolis.

Anderson, A.B., May, P.H., and Balick, M.J., 1991. *The subsidy from nature: palm forests, peasantry, and development on an Amazon frontier*. New York: Columbia University Press.

Anderson, B., 1991. *Imagined communities: reflections on the origin and spread of nationalism*. London: Verso.

Balée, W.L., 1994. *Footprints of the forest: Ka'apor ethnobotany-the historical ecology of plant utilization by an Amazonian people*. New York: Columbia University Press.

Becker, B.K., 1982. *Geopolítica da Amazônia: a nova fronteira de recursos*. Rio de Janeiro: Zahar Editores.

Bicker, A., Sillitoe, P., and Pottier, J., 2003. *Negotiating local knowledge: power and identity in development*. Sterling, Virginia: Pluto Press.

Birkner, W.M.K., 2002. *O realismo de Golbery: segurança nacional e desenvolvimento global no pensamento de Golbery do Couto e Silva*. Itajaí: Univali.

Block, D., 1994. *Mission culture on the upper Amazon: native tradition, Jesuit enterprise & secular policy in Moxos, 1660–1880*. Lincoln: University of Nebraska Press.

Bolaños, O., forthcoming 2011. Redefining identities, redefining landscapes: indigenous identity and land rights struggles in the Brazilian Amazon. *Journal of Cultural Geography.*

Borner, J. and Wunder, S., 2008. Paying for avoided deforestation in the Brazilian Amazon: from cost assessment to scheme design. *International Forestry Review,* 10, 496–511.

Borner, J. and Wunder, S., 2010. Direct conservation payments in the Brazilian Amazon: scope and equity implications. *Ecological Economics,* 69, 1272–1282.

Bowman, M.S., Amacher, G.S., and Merry, F.D., 2008. Fire use and prevention by traditional households in the Brazilian Amazon. *Ecological Economics,* 67, 117–130.

Branford, S. and Glock, O., 1985. *The last frontier: fighting over land in the Amazon.* London: Zed.

Brannstrom, C., 2009. South America's neoliberal agricultural frontiers: places of environmental sacrifice or conservation opportunity? *Ambio,* 38, 141–149.

Brass, T., 2002. Latin American peasants-new paradigms for old? *Journal of Peasant Studies,* 29, 1–40.

Brass, T., 2005. Neoliberalism and the rise of (peasant) nations within the nation: Chiapas in comparative and theoretical perspective. *Journal of Peasant Studies,* 32, 651–691.

Brondizio, E.S., 2008. *The Amazon caboclo and the acai palm: forest farmers in the global market.* The Bronx, NY: New York Botanical Garden.

Brondizio, E.S., Ostrom, E., and Young, O.R., 2009. Connectivity and the governance of multilevel social-ecological systems: the role of social capital. *Annual Review of Environment and Resources,* 34, 253–278.

Brondizio, E.S., Siqueira, A.D. and Vogt, N., forthcoming 2011. Forest resources, city services: globalization, household networks, and urbanization in the Amazon estuary. *In*: S.B. Hecht, C. Padoch and K. Morrison, eds. *The social lives of forests.* Chicago: University of Chicago Press.

Brown, K. and Rosendo, S., 2000. The institutional architecture of extractive reserves in Rondonia, Brazil. *Geographical Journal,* 166, 35–48.

Campos, M.T. and Nepstad, D.C., 2006. Smallholders, the Amazon's new conservationists. *Conservation Biology,* 20, 1553–1556.

Chasteen, J.C. and Castro-Klarén, S., 2003. *Beyond imagined communities: reading and writing the nation in nineteenth-century Latin America.* Washington, DC: Woodrow Wilson Center Press.

Chernilo, D., 2007. *A social theory of the nation-state: the political forms of modernity beyond methodological nationalism.* London: Routledge.

Chatre, A. and Agrawal, A., 2009. Trade-offs and synergies between carbon storage and livelihood benefits from forest commons. *Proceedings of the National Academy of Sciences of the United States of America,* 106, 17667–17670.

Cleary, D., 2001. Towards an environmental history of the Amazon: from prehistory to the nineteenth century. *Latin American Research Review,* 36, 65–96.

Dandy, N., 2005. Extractive reserves in Brazilian Amazonia: local resource management and the global political economy. *Environmental Politics,* 14, 434–436.

Davis, S.H., 1977. *Victims of the miracle: development and the Indians of Brazil.* Cambridge; New York: Cambridge University Press.

Descola, P., 1994. *In the society of nature: a native ecology in Amazonia.* Cambridge: Cambridge University Press.

Descola, P., 1996. *The spears of twilight: life and death in the Amazon jungle.* London: HarperCollins.

Escobar, A., 2008. *Territories of difference: place, movements, life, redes.* Durham: Duke University Press.

Fausto, C., 2001. *Inimigos fiéis: história, guerra e xamanismo na Amazônia.* São Paulo, SP, Brasil: Edusp.

Fausto, C. and Heckenberger, M., 2007. *Time and memory in indigenous Amazonia: anthropological perspectives.* Gainesville: University Press of Florida.

Fearnside, P.M., 2007. Brazil's Cuibá-Santarém (BR-163) Highway: the environmental cost of paving a soybean corridor through the Amazon. *Environmental Management*, 39, 601–614.

Fearnside, P.M., 2008. The roles and movements of actors in the deforestation of Brazilian Amazonia. *Ecology and Society*, 13 (1), 23. Available from: http://www.ecologyandsociety.org/vol13/iss1/art23/ [Accessed 20 October 2010].

Foresta, R.A., 1991. *Amazon conservation in the age of development: the limits of providence.* Gainesville: University of Florida Press Center for Latin American Studies.

French, J.H., 2006. Buried alive: imagining Africa in the Brazilian Northeast. *American Ethnologist*, 33, 340–360.

Garfield, S., 2001. *Indigenous struggle at the heart of Brazil.* Durham: Duke.

Goeschl, T. and Igliori, D.C., 2004. Reconciling conservation and development: a dynamic hotelling model of extractive reserves. *Land Economics*, 80, 340–354.

Gomes, F.d.S., 2005. *A hidra e os pântanos: mocambos, quilombos e comunidades de fugitivos no Brasil (séculos XVII-XIX).* São Paulo: Unesp.

Grainger, A., *et al.*, 2009. Biodiversity and REDD at Copenhagen. *Current Biology*, 19, R974–R976.

Hall, A.L., 1997. *Sustaining Amazonia: grassroots action for productive conservation.* Manchester, UK: Manchester University Press.

Hall, A.L., 2000. *Amazonia at the crossroads: the challenge of sustainable development.* London: Institute of Latin American Studies.

Harris, M., 2010. *Rebellion on the Amazon: the Cabanagem, race and popular culture in the north of Brazil 1798–1840.* New York: Cambridge Press.

Harris, M. and Nugent, S., 2004. *Some other Amazonians: perspectives on modern Amazonia.* London: Institute for the Study of the Americas.

Hecht, S.B., 2005. Soybeans, development and conservation on the Amazon frontier. *Development and Change*, 36, 375–404.

Hecht, S.B., 2010. The new rurality: globalization, peasants and the paradoxes of landscapes. *Land Use Policy*, 27, 161–169.

Hecht, S.B., forthcoming 2011a. Multiple environmentalisms: politics, interdisciplinarities, political ecologies and the decline of deforestation in Amazonia. *Environmental Conservation.*

Hecht, S.B., forthcoming 2011b. *The scramble for the Amazon: imperial contests and the tropical odyssey of Euclides da Cunha.* Chicago: University of Chicago Press.

Hecht, S.B. and Cockburn, A., 1989. *The fate of the forest: developers, destroyers, and defenders of the Amazon*. London: Verso.

Hecht, S.B. and Mann, C.C., 2008. How Brazil outfarmed the American farmer. *Fortune*, 157, 92–105.

Hemming, J., 1987. *Amazon frontier: the defeat of the Brazilian Indians*. London: Macmillan.

Hemming, J., 2003. *Die if you must: Brazilian Indians in the twentieth century*. London: Macmillan.

Hill, J.D., 1988. *Rethinking history and myth: indigenous South American perspectives on the past*. Urbana: University of Illinois Press.

Hoch, L., Pokorny, B., and De Jong, W., 2009. How successful is tree growing for smallholders in the Amazon? *International Forestry Review*, 11, 299–310.

Hochstetler, K. and Keck, M., 2007. *Greening Brazil: environmental activism in state and society*. Durham: Duke University Press.

Holston, J., 1989. *The modernist city: an anthropological critique of Brasília*. Chicago: University of Chicago Press.

Holston, J., 1991. The misrule of law-land and usurpation in Brazil. *Comparative Studies in Society and History*, 33, 695–725.

Holston, J., 1999. Alternative modernities: statecraft and religious imagination in the Valley of the Dawn. *American Ethnologist*, 26, 605–631.

Holston, J., 2008. *Insurgent citizenship: disjunctions of democracy and modernity in Brazil*. Princeton: Princeton University Press.

Holston, J. and Appadurai, A., 1996. Cities and citizenship. *Public Culture*, 8, 187–204.

INPE and Espacias, I.N.d.P., 2010. Desmatemento Cai.

Jepson, W., 2006. Producing a modern agricultural frontier: firms and cooperatives in Eastern Mato Grosso, Brazil. *Economic Geography*, 82, 289–316.

Jepson, W.E., Brannstrom, C., and de Souza, R.S., 2005. A case of contested ecological modernisation: the governance of genetically modified crops in Brazil. *Environment and Planning C-Government and Policy*, 23, 295–310.

Joseph, G.M. and Nugent, D., 1994. *Everyday forms of state formation: revolution and the negotiation of rule in modern Mexico*. Durham: Duke University Press.

Kainer, K., *et al.*, 2003. Experiments in forest-based development in Western Amazonia. *Society & Natural Resources*, 16, 869–886.

Kotscho, R., 1981. *O massacre dos posseiros: conflitos de terras no Araguaia-Tocantins*. Säao Paulo: Editora Brasiliense.

Lima, E., *et al.*, 2006. Searching for sustainability: forest policies, smallholders, and the trans-Amazon highway. *Environment*, 48, 36.

Mallon, F., 1995. *Peasant and nation: the making of post colonial Mexico and Peru*. Berkeley: Univeristy of California Press.

Martins, J.D., 2002. Representing the peasantry? Struggles for/about land in Brazil. *Journal of Peasant Studies*, 29, 300–335.

Martins, J.R. and Zirker, D., 2000. Nationalism, national security, and Amazonia: military perceptions and attitudes in contemporary Brazil. *Armed Forces & Society*, 27, 105–129.

Mattos, C.d.M., 1980. *Uma geopolítica pan-amazônica*. Rio de Janeiro: José Olympio Editora em convênio com o Instituto Nacional do Livro.

Maxwell, K., 1995. *Pombal, paradox of the Enlightenment*. Cambridge: Cambridge University Press.

Maxwell, K., 2001. The spark: Pombal, the Amazon, and the Jesuits. *Portuguese Studies*, 17, 168–183.

Medina, G., Pokorny, B., and Campbell, B., 2009a. Loggers, development agents and the exercise of power in Amazonia. *Development and Change*, 40, 745–767.

Medina, G., Pokorny, B., and Campbell, B., 2009b. Community forest management for timber extraction in the Amazon frontier. *International Forestry Review*, 11, 408–420.

Mendes, C., Gross, T. and Latin America Bureau, 1992. *Fight for the forest: Chico Mendes in his own words*. London: Latin America Bureau.

Mentore, L.H., forthcoming 2011. Waiwai fractality and the arboreal bias of PES schemes in Guyana: what to make of the multiplicity of Amazonian cosmographies? *Journal of Cultural Geography*.

Merry, F., Amacher, G., and Lima, E., 2008. Land values in frontier settlements of the Brazilian Amazon. *World Development*, 36, 2390–2401.

Merry, F., *et al.*, 2006. Collective action without collective ownership: community associations and logging on the Amazon frontier. *International Forestry Review*, 8, 211–221.

Michalski, F., Peres, C.A., and Lake, I.R., 2008. Deforestation dynamics in a fragmented region of southern Amazonia: evaluation and future scenarios. *Environmental Conservation*, 35, 93–103.

Minzenberg, E. and Wallace, R., forthcoming 2011. Amazonian agriculturalists bound by subsistence hunting. *Journal of Cultural Geography*.

Moran, E.F., 1993. *Through Amazonian eyes: the human ecology of Amazonian populations*. Iowa City: University of Iowa Press.

Nepstad, D., *et al.*, 2006a. Inhibition of Amazon deforestation and fire by parks and indigenous lands. *Conservation Biology*, 20, 65–73.

Nepstad, D.C., Stickler, C.M., and Almeida, O.T., 2006b. Globalization of the Amazon soy and beef industries: opportunities for conservation. *Conservation Biology*, 20, 1595–1603.

Nugent, S., 1993. *Amazonian caboclo society: an essay on invisibility and peasant economy*. Providence, RI: Berg.

Nygren, A., 2000. Development discourses and peasant-forest relations: natural resource utilization as social process. *Development and Change*, 31, 11–34.

Otero, G. and Jugenitz, H.A., 2003. Challenging national borders from within: the political-class formation of indigenous peasants in Latin America. *Canadian Review of Sociology and Anthropology-Revue canadienne de sociologie et d'anthropologie*, 40, 503–524.

Padoch, C., 1999. *Várzea: diversity, development, and conservation of Amazonia's whitewater floodplains*. Bronx, NY: New York Botanical Garden Press.

Padoch, C., *et al.*, 2008. Urban forest and rural cities: multi-sited households, consumption patterns, and forest resources in Amazonia. *Ecology and Society* 13 (2), 2. Available from: http://www.ecologyandsociety.org/vol13/iss2/art2/ [Accessed 2 November 2010].

Padoch, C., *et al.*, forthcoming 2011. Urban residence, rural employment, and the future of Amazonian forests. *In*: S.B. Hecht, C. Padoch and K. Morrison, eds. *The Social Lives of Forests*. Chicago: University of Chicago Press.

Parry, L., *et al.*, 2010. Rural-urban migration brings conservation threats and opportunities to Amazonian watersheds. *Conservation Letters*, 3, 251–259.

Pedlowski, M.A., *et al.*, 2005. Conservation units: a new deforestation frontier in the Amazonian state of Rondonia, Brazil. *Environmental Conservation*, 32, 149–155.

Peres, C.A. and Zimmerman, B., 2001. Perils in parks or parks in peril? Reconciling conservation in Amazonian reserves with and without use. *Conservation Biology*, 15, 793–797.

Perrault, T., 2008. Popular protest and unpopular policies: state restructuring, resource conflict and social justice in Bolivia. *In*: D. Carruthers, ed. *Environmental justice in Latin America*. Cambridge: MIT Press, 239–262.

Perreault, T., 2003. Social capital, development, and indigenous politics in Ecuadorian Amazonia. *Geographical Review*, 93, 328–349.

Perz, S.G., *et al.*, 2010. Intraregional migration, direct action land reform, and new land settlements in the Brazilian Amazon. *Bulletin of Latin American Research*, 29, 459–476.

Perz, S.G., Warren, J., and Kennedy, D.P., 2008. Contributions of racial-ethnic reclassification and demographic processes to indigenous population resurgence-the case of Brazil. *Latin American Research Review*, 43, 7–33.

Pieck, S.K., forthcoming 2011. Beyond postdevelopment: civic responses to regional integration in the Amazon. *Journal of Cultural Geography*.

Pinedo-Vasquez, M. and Padoch, C., 2009. Urban and rural and in-between: multi-sited households, mobility and resource management in the Amazon floodplain. *In*: M. Alexiades, ed. *Mobility and migration in indigenous Amazonia: contemporary ethnoecological perspectives*. Oxford: Berghahn.

Porro, A., 1996. *O povo das águas: ensaios de etno-história amazônica*. Petrópolis; São Paulo, SP: Vozes; Edusp.

Porro, N., Veiga, I. and Mota, D., forthcoming 2011. Traditional communities in the Brazilian Amazon and the emergence of new political identities: the struggle of the *quebradeiras de coco babaçu*-Babassu breaker women. *Journal of Cultural Geography*.

Posey, D.A. and Plenderleith, K., 2004. *Indigenous knowledge and ethics: a Darrell Posey reader*. New York: Routledge.

Raffles, H., 2002. *In Amazonia: a natural history*. Princeton, NJ: Princeton University Press.

Raffles, H. and WinklerPrins, A., 2003. Further reflections on Amazonian environmental history: transformations of rivers and streams. *Latin American Research Review*, 38, 165–187.

Reed, J.P., 2003. Indigenous land policies, culture and resistance in Latin America. *Journal of Peasant Studies*, 31, 137–156.

Restall, M., 2005. *Beyond black and red: African-native relations in colonial Latin America*. Albuquerque: University of New Mexico Press.

Rival, L., 2003. From peasant struggles to Indian resistance: the Ecuadorian Andes in the late twentieth century. *Journal of Peasant Studies*, 30, 217–220.

Rival, L., 2004. Amazonia: territorial struggles on perennial frontiers. *Journal of Latin American Studies*, 36, 603–605.

Rival, L.M., 2002. *Trekking through history: the Huaorani of Amazonian Ecuador*. New York: Columbia University Press.

Rodrigues, G. and Rabben, L., 2007. *Walking the forest with Chico Mendes: struggle for justice in the Amazon*. Austin: University of Texas Press.

Rostow, W.W., 1971. *The stages of economic growth; a non-communist manifesto.* Cambridge: Cambridge University Press.

Ruiz-Perez, M., *et al.*, 2005. Conservation and development in Amazonian extractive reserves: the case of Alto Jurua. *Ambio*, 34, 218–223.

Salisbury, D.S., Borgo López, J. and Vela Alvarado, J.W., forthcoming 2011. Transboundary political ecology in Amazonia: history, culture, and conflicts of the borderland Asháninka. *Journal of Cultural Geography.*

Sandbrook, C., *et al.*, 2010. Carbon, forests and the REDD paradox. *Oryx*, 44, 330–334.

Schelling, V., 2001. *Through the kaleidoscope: the experience of modernity in Latin America.* London: Verso.

Schmink, M. and Wood, C.H., 1992. *Contested frontiers in Amazonia.* New York: Columbia University Press.

Scott, J.C., 1998. *Seeing like a state.* New Haven: Yale University Press.

Sears, R.R., Padoch, C., and Pinedo-Vasquez, M., 2007. Amazon forestry tranformed: integrating knowledge for smallholder timber managemet in eastern Brazil. *Human Ecology*, 35, 697–707.

Segebart, D., 2008. Who governs the Amazon? Analysing governance in processes of fragmenting development: policy networks and governmentality in the Brazilian Amazon. *Erde*, 139, 187–205.

Sikor, T., *et al.*, 2010. REDD-plus, forest people's rights and nested climate governance. *Global Environmental Change-Human and Policy Dimensions*, 20, 423–425.

Silva, G.d.C.e., 1967. *Geopolítica do Brasil.* Rio de Janeiro: J. Olympio.

Silva, G.d.C.e., 2003. *Geopolitica e poder.* Rio de Janeiro: UniverCidade.

Simmons, C.S., *et al.*, 2007. The Amazon land war in the south of Pará. *Annals of the Association of American Geographers*, 97, 567–592.

Stoian, D., 2005. Making the best of two worlds: rural and peri-urban livelihood options sustained by nontimber forest products from the Bolivian Amazon. *World Development*, 33, 1473–1490.

Taleb, N., 2009. *The black swan: the impact of the highly improbable.* New York: Random House.

Tocantins, L., 1966. Arquitetura e paisagismo na Amazônia. Manaus: Govêrno do Estado do Amazonas, Secretaria de Imprensa e Divulga Ção.

Toni, F., 2007. *Expansão e trajetórias da pecuária na Amazônia: Acre, Brasil.* Brasília, DF: Editora UnB.

Treccani, G., 2007. *Territórios quilombolas.* Belém, Brasil: INTERPA.

Turner, B.L. and Robbins, P., 2008. Land-change science and political ecology: similarities, differences, and implications for sustainability science. *Annual Review of Environment and Resources*, 33, 295–316.

Vadjunec, J.M., Schmink, M. and Gomes, C.V.A., forthcoming 2011. Rubber tappper citizens: emerging places, policies, and shifting identities in Acre, Brazil. *Journal of Cultural Geography.*

Vargas, G., 1938. *A nova politica do Brasil.* Rio de Janeiro: Jose Olimpio.

Viana, V., 2009. Seeing REDD in the Amazon. London: IIED.

Walker, R., *et al.*, 2009. Ranching and the new global range: Amazonia in the 21st century. *Geoforum*, 40, 732–745.

Wood, C.H. and Schmink, M., 1993. The military and the environment in the Brazilian Amazon. *Journal of Political & Military Sociology*, 21, 81–105.

Wright, R., 1998. *Cosmos, self, and history in Baniwa religion: for those unborn.* Austin: University of Texas Press.

Zimmerer, K.S., 2000. The reworking of conservation geographies: nonequilibrium landscapes and nature-society hybrids. *Annals of the Association of American Geographers*, 90, 356–369.

Zimmerer, K.S., 2006. Cultural ecology: at the interface with political ecology-the new geographies of environmental conservation and globalization. *Progress in Human Geography*, 30, 63–78.

Index